I0489272

NUREG-1022
Rev. 1

Event Reporting Guidelines
10 CFR 50.72 and 50.73

Manuscript Completed: December 1997
Date Published: January 1998

Prepared by
D. P. Allison, M. R. Harper, W. R. Jones,
J. B. MacKinnon, S. S. Sandin

Office for Analysis and Evaluation of Operational Data
U.S. Nuclear Regulatory Commission
Washington, DC 20555-0001

Impact

This Revision 1 to NUREG-1022 clarifies and consolidates the guidance on implementing the event notification and reporting requirements in 10 CFR 50.72 and 50.73. Little of the guidance is new or different from the generic reporting guidance previously published in final form in NUREG-1022 (1983), its Supplement 1 (1984) and subsequent generic communications. Where it is different, the changes are minor. In some areas the new guidance will result in fewer reports and in some areas it will result in more reports. On balance, the clarified guidance will result in a small decrease in reporting burden.

PAPERWORK REDUCTION ACT STATEMENT

This report amends the information collections contained in 10 Code of Federal Regulations (CFR) Part 50 and NRC Form 366, Licensee Events Reports. The changes are considered to be insignificant when compared with the overall requirements of the CFR Part and the form (NRC Form 366 reduction of 350 hours annually vs. the current 75K, and 10 CFR 50.72 reduction of 150 hours annually vs. the current 2.4K). NRC does not consider the burden to be significant enough to trigger the requirements of the Paperwork Reduction Act of 1995 (44 U.S.C. 3501 et seq.). Existing requirements were approved by the Office of Management and Budget (OMB), approval number 3150-0011 and 3150-0104.

Public Protection Notification

If an information collection does not display a currently valid OMB control number, the NRC may not conduct or sponsor, and a person is not required to respond to, the information collection.

ABSTRACT

Revision 1 to NUREG-1022 clarifies the immediate notification requirements of Title 10 of the *Code of Federal Regulations*, Part 50, Section 50.72 (10 CFR 50.72), and the 30-day written licensee event report (LER) requirements of 10 CFR 50.73 for nuclear power plants. This revision was initiated to improve the reporting guidelines related to 10 CFR 50.72 and 50.73 and to consolidate these guidelines into a single reference document. A first draft of this document was noticed for public comment in the *Federal Register* on October 7, 1991 (56 FR 50598). A second draft was noticed for comment in the *Federal Register* on February 7, 1994 (59 FR 5614). This document updates and supersedes NUREG-1022 and its Supplements 1 and 2 (published in September 1983, February 1984, and September 1985, respectively). It does not change the reporting requirements of 10 CFR 50.72 and 50.73.

CONTENTS

APPENDICES

TABLES

EXECUTIVE SUMMARY

Two of the many elements contributing to the safety of nuclear power are emergency response and the feedback of operating experience into plant operations. These are achieved partly by the licensee event reporting requirements of Title 10 of the *Code of Federal Regulations*, Part 50, Sections 50.72 and 50.73 (10 CFR 50.72 and 50.73), which became effective on January 1, 1984. Section 50.72 provides for immediate notification requirements via the emergency notification system (ENS) and Section 50.73 provides for 30-day written licensee event reports (LERs).

The information reported under 10 CFR 50.72 and 50.73 is used by the NRC staff in responding to emergencies, monitoring ongoing events, confirming licensing bases, studying potentially generic safety problems, assessing trends and patterns of operational experience, monitoring performance, identifying precursors of more significant events, and providing operational experience to the industry.

Experience has shown that the threshold for reporting has not been consistently implemented and some problems exist with the interpretation of the guidelines and definitions. A 1990 survey on the effect of NRC regulation on nuclear power plant activities and subsequent event reporting workshops also indicated a need for further guidance on the two reporting rules.

Therefore, the NRC staff prepared NUREG-1022, Revision 1, which clarifies implementation of the existing 10 CFR 50.72 and 50.73 rules and consolidates important NRC reporting guidelines into one reference document. The clarifications include major editing of the previous guidelines. The document is structured to assist licensees in achieving prompt and complete reporting of specified events and conditions. The revised guidelines are not expected to result in a significant change in the annual industry-wide total numbers for ENS notifications and LERs. The effect on individual licensees is expected to vary.

The document addresses general issues of reporting that have not been consistently applied and covers such diverse subjects as engineering judgment, multiple failures and related events, deficiencies discovered during licensee engineering reviews, and human performance issues. The guidelines for specific reporting criteria have been enhanced by improved discussions of concepts, thresholds, and illustrative examples; definitions of key terms and phrases; and original ENS guidelines for some criteria that were not previously addressed. A new section has been added that discusses ENS communications and methods, voluntary reporting, retraction of reports, importance of reporting timeliness and completeness, and typical NRC concerns associated with ENS notifications for each reporting requirement.

ACKNOWLEDGMENTS

The authors Revision 1 wish to acknowledge the members of the task group that prepared the first draft of Revision 1: John Boardman, Paul Bobe, Marcel Harper, John Crooks, L. Mark Padovan, and Robert Spence of the U.S. Nuclear Regulatory Commission (NRC) Office for Evaluation and Operation of Operational Data (AEOD). They were assisted by Roger Woodruff of the NRC Office of Nuclear Reactor Regulation (NRR) and Eric Weiss and Aimee Brown of AEOD. With regard to the second draft and this final version, the authors wish to acknowledge the contributions of Sanford Israel and Shirley Rohrer of AEOD. With regard to the final draft the authors also wish to acknowledge the assistance of David Skeen of NRR, Milton Vagins of the NRC Office of Nuclear Regulatory Research, and representatives of the NRC's four regional offices, Gordon Hunegs (RI), Charles Ogle (RII), Edward Schweibinz (RIII) and Michael Runyan (RIV). Dee Gable of the NRC Office of Administration (ADM) reviewed the first two drafts. Geary Mizuno of the NRC Office of the General Counsel (OGC) reviewed both drafts and the final version.

ABBREVIATIONS

AEOD	Analysis and Evaluation of Operational Data, Office for
AIT	augmented inspection team
ASME	American Society of Mechanical Engineers
ASP	accident sequence precursor
ATWS	anticipated transient without scram
BPV	Boiler and Pressure Vessel Code (ASME)
BWR	boiling-water reactor
CFR	*Code of Federal Regulations*
CRDM	control rod drive mechanism
CRVS	control room ventilation system
DBDR	design-basis documentation review
DDR	design document reconstitution
ECCS	emergency core cooling system
EDG	emergency diesel generator
EIIS	Energy Industry Identification System
ENS	emergency notification system
EO	emergency officer
EOF	emergency operations facility
EOP	emergency operating procedure
EPIX	equipment performance and information exchange
EPA	Environmental Protection Agency (U.S.)
ERDS	emergency response data system
ERF	emergency response facility
ESF	engineered safety feature(s)
ESW	emergency service water
FEMA	Federal Emergency Management Agency
FFD	fitness for duty
FSAR	final safety analysis report
FTS	federal telecommunications system
GDC	general design criteria
GL	generic letter
HOO	headquarters operations officer

HP	health physics
HPCI	high-pressure coolant injection
HPI	high-pressure injection
HPN	health physics network
HPSI	high pressure safety injection
HVAC	heating, ventilation and air conditioning
IEEE	Institute of Electrical and Electronics Engineers
IIT	incident investigation team
ILRT	integrated leak rate test
IN	information notice
INPO	Institute of Nuclear Power Operations
ISI	inservice inspection
IST	inservice testing
ISTS	improved standard technical specifications
LCO	limiting condition for operation
LER	licensee event report
LOCA	loss of coolant accident
LPSW	low pressure service water
MPC	maximum permissible concentration
MSIV	main steam isolation valve
NPRDS	nuclear plant reliability data system
NRC	Nuclear Regulatory Commission (U.S.)
NRR	Nuclear Reactor Regulation, Office of
NUMARC	Nuclear Management and Resources Council
OCR	optical character reader
OMS	overpressure mitigation system
PDR	Public Document Room
PGA	policies, guidance, and administrative controls
PWR	pressurized water reactor
RAB	Reactor Analysis Branch
RBVS	reactor building ventilation system
RCIC	reactor core isolation cooling
RCP	reactor coolant pump
RCS	reactor coolant system
RDO	regional duty officer
RHR	residual heat removal
RPS	reactor protection system
RWCU	reactor water cleanup
SALP	systematic assessment of licensee performance

SAR	safety analysis report
S/D	shutdown
SIS	safety injection system
SOV	solenoid-operated valve
SPDS	safety parameter display system
SRO	senior reactor operator
STS	standard technical specifications
TS	technical specification(s)
TSC	technical support center

1 INTRODUCTION

This document provides guidance on the reporting requirements of Title 10 of the *Code of Federal Regulations*, Part 50, Sections 50.72 and 50.73 (10 CFR 50.72 and 10 CFR 50.73). While these reporting requirements range from immediate, 1-hour, and 4-hour verbal notifications to 30-day written reports, covering a broad spectrum of events from emergencies to generic component level deficiencies, the NRC wishes to emphasize that reporting requirements should not interfere with ensuring the safe operation of a nuclear power plant. Licensees' immediate attention must always be given to operational safety concerns.

1.1 Background

The origins of 10 CFR 50.72 and 50.73 are described in Appendix A to this report. In 1983, partially in response to lessons from the Three Mile Island accident, the U.S. Nuclear Regulatory Commission (NRC) revised its immediate notification requirements via the emergency notification system (ENS) in 10 CFR 50.72 and modified and codified its written licensee event report (LER) system requirements in 10 CFR 50.73. The revision of 10 CFR 50.72 and the new 10 CFR 50.73 became effective on January 1, 1984. Together, they specify the types of events and conditions reportable to the NRC for emergency response and identifying plant-specific and generic safety issues.

The two rules have identical reporting thresholds and similar language whenever possible. They are complementary and of equal importance, with necessary dissimilarities in reporting requirements to meet their different purposes, as illustrated in this report, Section 1, Table 1, and Section 3 text.

Section 50.72 is structured to provide telephone notification of reportable events to the NRC Operations Center within a time frame established by the relative importance of the events. Events are categorized as either emergencies (immediate notifications, but no later than 1 hour) or non emergencies. The latter is further categorized into 1-hour and 4-hour notifications; non-emergency events requiring 4-hour notifications generally have slightly less urgency and safety significance than those requiring 1-hour notifications. Immediate telephone notification to the NRC Operations Center of declared emergencies is necessary so the Commission may immediately respond. Reporting of non-emergency events and conditions is necessary to permit timely NRC followup via event monitoring, special inspections, generic communications, or resolution of public or media concerns.

Section 50.73 requires written LERs to be submitted on reportable events within 30 days of their occurrence, after a thorough analysis of the event, its root causes, safety assessments, and corrective actions are available, to permit NRC engineering analyses and studies.

Some reporting guidance for 10 CFR 50.72 and 50.73 was contained in the Statements of Considerations for the rules. More detailed guidelines and examples of reportable events were developed and issued in NUREG-1022 and its Supplements 1 and 2. The intent of these publications was to achieve complete reporting of specified events and conditions. Subsequently, additional interpretations and directions on certain subjects have been issued in NRC bulletins, information notices, and generic letters.

1.2 Reporting Guidelines and Industry Experience

Event reporting under these rules since 1984 has contributed significantly to focusing the attention of the NRC and the nuclear industry on the lessons learned from operating experience to improve reactor safety. In the mid-1980's, decreasing trends in the number of reactor transients and in the number of significant events and improvements in reactor safety system performance were noticeable. Since 1989, these trends have leveled off as fewer plants were on a learning curve and industry completed improvements that have a high return in safety performance. While the more obvious lessons have been extracted from operating experience, more analyses need to be performed and new efforts need to be developed to extract further lessons from operational data.

The operational experience submitted in accordance with 10 CFR 50.72 and 50.73 is publicly available and has been used by other organizations in ways that are most often beneficial to nuclear safety. However, uses in areas that were unintended, such as in prudence hearings, in statistical presentations and comparisons of reporting rates without regard to or inclusion of a technical analysis of the safety significance of the events, can lead to unwarranted impressions of safety performance. In such uses, there has been a tendency to only count the number of reported events without assessing their individual safety significance. Such misuses could result in licensees adopting a more restrictive reporting threshold in order to reduce the number of reportable events, although the Commission's requirement for a low threshold has not changed. This can be counterproductive to the purpose of these rules.

Experience has shown that the threshold of reporting, as well as other areas of the reporting rules, has not been consistently implemented. Some problems have occurred in such areas as interpretation of the guidelines and definitions, timeliness of reporting, reporting of generic concerns, engineering judgment, and reporting of deficiencies found during design reviews. These problems, as well as a 1990 survey on the effect of NRC regulation on nuclear power plant activities and subsequent event reporting workshops, identified the need for further guidelines on the two reporting rules.

1.3 Revised Reporting Guidelines

The purpose of this revision to NUREG-1022 is to ensure events are reported as required by improving 10 CFR 50.72 and 50.73 reporting guidelines and to consolidate these guidelines into a single reference document. This document updates and supersedes NUREG-1022 and its Supplements 1 and 2.

An NRC task group prepared this document principally by editing and combining the information contained in NUREG-1022 and its Supplements 1 and 2, the Statements of Considerations for

10 CFR 50.72 and 50.73, other NRC staff documents on event reporting (such as information notices, bulletins, inspection manual chapters, enforcement actions, letters and memoranda), ENS event notification reports, and LERs. A second task group prepared the second draft of this document, principally by considering the public comments received and the requirements of the rules, their Statements of Considerations, and previous NRC generic guidance on reporting.

In compiling this document, the information in NUREG-1022 was edited for clarity. The paragraph-by-paragraph explanation of the LER rule, which was a restatement of guidance in the Statement of Considerations, was preserved or more thoroughly discussed. Most of the examples were replaced with others that have been condensed to exemplify specific reporting thresholds.

Most of the specific questions and answers on both rules as contained in NUREG-1022, Supplement 1, were incorporated as generic statements into the discussions or examples in Sections 2, 3, 4, and 5 of this document. The ENS and LER rules are compared side-by-side in Section 3.

NUREG-1022, Supplement 2, made recommendations for improvements in LER quality; Appendices B and D of Supplement 2 were incorporated into the discussions in Section 5.2 of this document.

In addition, experience from responding to NRC staff and licensee inquiries in various event reporting workshops since 1984 and ENS calls has been considered in this report. Many actual events were summarized to exemplify event reportability in response to licensee requests. The principal NRC staff involved in the original codification and revisions to 10 CFR 50.72 and 50.73 were consulted regarding the original intent of the regulations.

Section 2 clarifies specific areas of 10 CFR 50.72 and 50.73 that are applicable to many reporting criteria or that historically appear to be subject to varied interpretations. It covers such diverse subjects as engineering judgment, differences in tenses between the two rules, retraction and voluntary reporting, legal reporting requirements, and human performance issues.

Section 3 contains guidelines on event reporting for specific criteria in both rules by means of discussions and examples of reported events. To minimize repetition, similar criteria from both rules are addressed together. The format follows the order of 10 CFR 50.72 with 50.73 appropriately interwoven.

Section 3.1 addresses general methods of ENS reporting for declared emergencies and non-emergencies. Practical guidelines are given on making ENS emergency notifications. Requirements for LER reporting regardless of plant mode, power level, or the significance of an initiating item are specified.

Section 3.2 addresses ENS 1-hour reporting criteria and 30-day LERs. The existing ENS and LER guidelines related to plant shutdowns required by technical specification (TS), TS deviations per §50.54(x), and TS prohibited operations or conditions are reiterated. Plant operation in a degraded or unanalyzed condition, or outside the plant's operating and emergency procedures, is clarified by definitions and examples. The timing of ENS reporting of anticipated natural phenomenon or conditions threatening plant safety is explained to ensure good communication between licensees and the NRC during developing situations. Valid emergency core cooling system (ECCS) discharges into the reactor coolant system are defined and invalid ECCS discharges are identified as reportable within 4 hours as an engineered safety feature (ESF) actuation. Additional guidelines and thresholds are given on the ENS reporting of the loss of emergency assessment, response, or communications. The intent of the reporting criteria on internal plant safety threats, including such examples as fire, toxic gas, or radiation releases, is explained to also include any other internal safety threat. Floods and spills are discussed as another typical threat to plant safety and the terms "threat" and "significant hampering of site personnel" are defined.

Section 3.3 addresses 4-hour ENS notifications and 30-day LERs. Examples are provided for degraded or unanalyzed conditions found while the plant is shut down. Engineered safety feature and reactor protection systems actuations are discussed. Anticipated transient without scram (ATWS) system actuations are addressed. The 1992 revisions to 10 CFR 50.72 and 50.73 that reduced the reporting of engineered safety features actuations are also discussed. Terms are defined regarding the reporting criteria for events or conditions that alone could have prevented fulfillment of the safety function required for shutdown of the reactor, removal of residual heat, release of radioactive material, or mitigation of the consequences of an accident. Single, common-mode, and multiple independent failures reportable under this criterion are discussed. The discussion of LER reporting of common-mode failures of independent safety system trains defines a number of terms and notes their importance as precursors. The existing ENS and LER guidelines related to airborne or liquid releases are restated. Guidelines are clarified on ENS reporting of a contaminated person requiring transport to an offsite medical facility. The basis and report timing for the ENS reporting criteria regarding news releases or other government notifications are explained, as necessary, so that the NRC can appropriately respond to media or government inquiries and thresholds for reporting are clarified. The recently issued ENS reporting criterion regarding spent fuel storage cask problems is included.

Section 3.4 addresses the requirements for immediate ENS followup notifications during the course of an event. The requirement, means, and methods to maintain continuous or periodic communication with the NRC during events, if so requested, are explained.

Section 4 explains ENS communications (from existing information notices), reporting timeliness and completeness, voluntary notifications, and retractions. Appropriate ENS emergency notification methods are described.

Section 5 reiterates previous guidelines on administrative requirements, preparation, and submittal of LERs. It specifies the information an LER should contain and provides steps to be followed in preparing an LER. It also includes an expanded human performance discussion to achieve ENS and LER content that examines both equipment and human performance.

Appendix A provides the history of 10 CFR 50.72 and 50.73, associated NRC workshops, and an NRC regulatory impact study, which was one of the factors leading to this document.

Appendix B discusses the key NRC ENS personnel, range of NRC responses to ENS notifications, and NRC event review.

Appendix C addresses the NRC LER analysis and evaluation programs and other uses of LERs nationally and internationally.

Appendix D contains 10 CFR 50.72 including its Statement of Considerations as published in the *Federal Register*.

Appendix E contains 10 CFR 50.73 including its Statement of Considerations as published in the *Federal Register*.

Appendix F contains 1992 revisions to 10 CFR 50.72 and 10 CFR 50.73 including the Statement of Considerations as published in the *Federal Register*.

1.4   How to Use These Guidelines

This NUREG was designed primarily as a reference to help licensees determine event reportability, make ENS notifications, and prepare and submit LERs.

* Reportability Determination

   The applicable 10 CFR 50.72 and 50.73 reporting criteria are identified in the Table of Contents of this report, as well as in the respective rules. Because these rules have overlapping reporting requirements, it is not unusual to find an event reportable under more than one criterion. A reportable event is to be reported under the most immediate reporting requirements.

   Generally, many events and conditions that require an ENS notification also require the submittal of an LER, as reflected by many of the rules' parallel reporting requirements. The reporting determination guidelines in Section 3 for both 10 CFR 50.72 and 50.73 are presented together wherever possible in the "Discussion" and "Example" explanations for each paragraph. The differences between the ENS and LER reporting requirements are underlined. The differences are discussed when they are important. Key terms are defined and important concepts are identified in the "Discussion" sections. Events used as examples may be reportable under other criteria but are usually only evaluated for reportability under the specific criteria they appear under. General issues, such as timeliness, can also be found in Section 2.

   Other reporting requirements applicable to operating reactors include 10 CFR 50.9, 20.2202, 20.2203, 50.36, 72.74, 72.216, 73.71, and Part 21. When reports are required under these regulations, some parts require the use of 10 CFR 50.72 and 50.73 notifications and written reports. Duplicate reporting is not required.

- ENS Notification

  Once an event has been determined to be reportable under 10 CFR 50.72, an ENS notification is to be made. The ENS notification time limit can be found under the applicable §50.72 criteria in Section 3; if more than one reporting criterion applies, the shortest time limit should be met. Guidelines on the information to be reported may be found in Section 4.3. Practical information regarding the actual telephone call can be found in Sections 4.1 and 4.3.

- LER Preparation and Submittal

  Once an event has been determined to be reportable under 10 CFR 50.73, an LER is to be prepared and submitted. Administrative requirements and guidelines for submitting LERs can be found in Section 5.1. The requirements and guidelines for the content of LERs can be found in Section 5.2.

1.5 New or Different Guidance

Reporting guidance that is considered to be new or different in a meaningful way, relative to previously published generic reporting guidance, is indicated by shading the appropriate text. Occasionally, text is marked by both redline and strikeout in order to show that specific items are being deleted.

1.6 Planned Future Actions

The NRC staff recognizes a need to revise the reporting requirements in 10 CFR 50.72 and 50.73 to better align them with the NRC's current needs, including the move toward risk-informed regulation and reporting of design basis issues. Accordingly, the staff plans to develop a rulemaking plan and proceed expeditiously with rulemaking to accomplish these ends. However, rulemaking will take time, whereas the work on this document has been completed. Thus, it is considered worthwhile to issue this revision to NUREG-1022 to provide improved guidance now. In the future, as rule changes are developed, new guidance will be developed concurrently.

Table 1 Comparability of 10 CFR 50.72 and 50.73 Criteria

Event or Condition	ENS notification as soon as practical and in all cases within 1 hour	ENS notification as soon as practical and in all cases within 4 hours	30-day LER	NUREG Sect.
EMERGENCY CLASS	Immediately after notification of State and local authorities, but no later than 1 hour after declaration of emergency class defined in licensee's emergency plan [50.72(a)(1),(a)(2),(a)(3) and (a)(4)]		Note-Although not specifically mentioned in 10 CFR 50.73, many emergency class events involve reportable situations	3.1.1
TECHNICAL SPECIFICATIONS (TS):				
Plant shutdown (S/D) required by TS	Initiation of S/D required by TS [50.72(b)(1)(i)(A)]		Completion of S/D required by TS [50.73(a)(2)(i)(A)]	3.2.1
TS prohibited operations or conditions			Operation or condition prohibited by TS [50.73(a)(2)(i)(B)]	3.2.2
TS deviation authorized by 50.54(x)	Deviation from TS authorized by 50.54(x) [50.72(b)(1)(i)(B)]		Criterion [50.73(a)(2)(i)(C)] same as ENS 1 hour	3.2.3
DEGRADED CONDITION; UNANALYZED CONDITION, OUTSIDE DESIGN BASIS, NOT COVERED BY PROCEDURES:				
Plant, including its principal safety barriers, seriously degraded	During operation, serious degradation of plant including its principal safety barriers [50.72(b)(1	Found while shut down; had it been found in operation, would have been seriously degraded [50.72(b)(2)(i)]	Either in operation or S/D, condition of plant, including principal safety barriers, seriously degraded [50.73(a)(2)(ii)]	3.2.4, 3.3.1

Table 1 (continued)

Event or Condition	ENS notification as soon as practical and in all cases within 1 hour	ENS notification as soon as practical and in all cases within 4 hours	30-day LER	NUREG Sect.
DEGRADED...(CONTINUED): Plant in unanalyzed condition significantly compromising plant safety	During operation, plant in unanalyzed condition, significantly compromising plant safety [50.72(b)(1)(ii)(A)]	Found while shut down; had it been found in operation, would have been unanalyzed condition that significantly compromises plant safety [50.72(b)(2)(I)]	Either in operation or S/D, unanalyzed condition significantly compromising plant safety [50.73(a)(2)(ii)(A)]	3.2.4, 3.3.1
Plant outside design basis of plant	During operation, plant in condition outside design basis [50.72(b)(1)(ii)(B)]		While either in operation or S/D, plant was in condition outside design basis [50.73(a)(2)(ii)(B)]	3.2.4, 3.3.1
Plant in condition not covered by operating and emergency procedures	During operation, plant in condition not covered by operating and emergency procedures [50.72(b)(1)(ii)(C)]		While either in operation or S/D, plant was in condition not covered by operating and emergency procedures [50.73(a)(2)(ii)(C)]	3.2.4, 3.3.1
EXTERNAL THREAT TO PLANT SAFETY	Any natural phenomenon or other external condition that poses an actual threat to the safety of the plant or significantly hampers site personnel in performance of duties necessary for its safe operation [50.72(b)(1)(iii)]		Criterion [50.73(a)(2)(iii)] same as ENS 1 hour	3.2.5
EMERGENCY CORE COOLING SYSTEM (ECCS) DISCHARGE; ACTUATION OF ANY ENGINEERED SAFETY (ESF)	A valid ECCS signal that results, or should have resulted, in ECCS discharge into the reactor coolant system [50.72(b)(1)(iv)]	Manual or automatic actuation of any ESF, including the reactor protection system (RPS), occurs and was not preplanned as part of a test or reactor operation [50.72(b)(2)(ii)]	Criterion [50.73(a)(2)(iv)] encompasses both ENS 1 hour and 4 hours	3.2.6, 3.3.2

Table 1 (continued)

Event or Condition	ENS notification as soon as practical and in all cases within 1 hour	ENS notification as soon as practical and in all cases within 4 hours	30-day LER	NUREG Sect.
EVENT THAT ALONE COULD HAVE PREVENTED FULFILLMENT OF A SAFETY FUNCTION		Event or condition alone would have prevented fulfillment of safety function of system needed for S/D of the reactor, maintenance of a safe S/D condition, residual heat removal (RHR), control of release of radioactive material, or mitigation of the consequences of an accident [50.72(b)(2)(iii)]	Criterion [50.73(a)(2)(v)] same as ENS 4 hours. Need not report individual component failures under this paragraph if redundant equipment in same system was operable and available [50.73(a)(2)(vi)]	3.3.3
COMMON CAUSE OR CONDITION RESULTING IN INDEPENDENT TRAINS OR CHANNELS BECOMING INOPERABLE			Single cause or condition caused inoperability of at least one independent train or channel in multiple systems or two independent trains and channels in a single system designed for safe S/D, RHR, radiation release control, or accident mitigation [50.73(a)(2)(vii)]	3.3.4
RADIOACTIVE RELEASES: Airborne radioactivity releases		Airborne radioactivity released to an unrestricted area exceeds 20x the concentration specified in 10 CFR 20, Appendix B, Table 2, averaged over 1 hour [50.72(b)(2)(iv)(A)]	Criterion [50.73(a)(2)(viii)(A)] same as ENS 4 hours.	3.3.5

Table 1 (continued)

Event or Condition	ENS notification as soon as practical and in all cases within 1 hour	ENS notification as soon as practical and in all cases within 4 hours	30-day LER	NUREG Sect.
RELEASES (CONTINUED): Liquid effluent releases		Liquid effluent released to an unrestricted area exceeds 20x the concentration specified in 10 CFR 20, Appendix B, Table 2, for all radionuclides except tritium and dissolved noble gases, averaged over 1 hour [50.72(b)(2)(iv)(B)].	Criterion [50.73(a)(2)(viii)(B)] same as ENS 4 hours.	3.3.5
INTERNAL THREAT TO PLANT SAFETY	Any event that poses an actual threat to the safety of the plant or significantly hampers site personnel in the conduct of safe operation [50.72(b)(1)(vi)]		Criterion [50.73(a)(2)(x)] same as ENS 1 hour	3.2.8
LOSS OF EMERGENCY ASSESSMENT, OFFSITE RESPONSE, OR COMMUNICATIONS CAPABILITY	A major loss of capability occurs for emergency assessment, offsite response, or communications [50.72(b)(1)(v)]			3.2.7
TRANSPORT OF CONTAMINATED PERSON TO OFFSITE MEDICAL FACILITY		A radioactively contaminated person is transported to an offsite medical facility [50.72(b)(2)(v)]		3.3.6

Table 1 (continued)

Event or Condition	ENS notification as soon as practical and in all cases within 1 hour	ENS notification as soon as practical and in all cases within 4 hours	30-day LER	NUREG Sect.
NEWS RELEASE/OTHER GOVERNMENT NOTIFICATIONS		A news release is planned or other government agencies have been or will be notified of an event related to the health and safety of the public or onsite personnel, or the protection of the environment [50.72(b)(2)(vi)]		3.3.7
DEGRADED SPENT FUEL STORAGE CASK OR CONFINEMENT SYSTEM		A defect in any spent fuel storage cask structure, system, or component that is important to safety [50.72(b)(2)(vii)(A)]. A significant reduction in the effectiveness of any spent fuel storage cask confinement system during use of the storage cask under a general licensee issued under 10 CFR 72.210 [50.72(b)(2)(vii)(B)]		3.3.8

Followup Notification (Section 3.4):
After making a 1-hour or 4-hour notification, licensees are required to immediately notify the NRC Operations Center if any of the following occur:

- plant conditions worsen [50.72(c)(1)(i)], emergency classification changed [50.72(c)(1)(ii)], or emergency class terminated [50.72(c)(1)(iii)];
- the results of ensuing evaluations or assessments of plant conditions are obtained [50.72(c)(2)(i)];
- the effectiveness of response or protective measures taken becomes known [50.72(c)(2)(ii)];
- information related to plant behavior is not understood [50.72(c)(2)(iii)];

In addition, if requested by the NRC, maintain an open, continuous communication channel with the NRC Operations Center [50.72(c)(3)].

NUREG-1022, Rev. 1

2 REPORTING AREAS WARRANTING SPECIAL MENTION

This section clarifies specific areas that are applicable to many reporting criteria or that historically appear to be subject to varied interpretations.

2.1 Engineering Judgment

The reportability of many events and conditions is self evident. However, the reportability of other events and conditions may not be readily apparent and the use of engineering judgment is involved in determining reportability.

Engineering judgment may include either a documented engineering analysis or a judgment by a technically qualified individual, depending on the complexity, seriousness, and nature of the event or condition. A documented engineering analysis is not a requirement for all events or conditions, but it would be appropriate for particularly complex situations. In addition, although not required by the rule, it may be prudent to record in writing that a judgment was exercised by identifying the individual making the judgment, the date made, and briefly documenting the basis for this judgment. In any case, the staff considers that the use of engineering judgment implies a logical thought process that supports the judgment.

2.2 Differences in Tense Between 10 CFR 50.72 and 50.73

The present tense was used in 10 CFR 50.72 because the event or condition generally would be ongoing at the time of reporting. The past tense was used in 10 CFR 50.73 because the event or condition normally would be past when an LER was written.

This difference creates some confusion over the reportability under 10 CFR 50.72 of events not related to an ongoing event or discovered as the result of an event review. In other cases, questions are raised regarding the need for a 10 CFR 50.73 report. Where the tense is relevant to reportability, it is addressed in the specific criterion in Section 3 of this report.

2.3 Multiple Failures and Related Events

More than one failure or event may be reported in a single ENS notification or LER if (1) the failures or events are related (have the same general cause or consequences) and (2) they occurred during a single activity (e.g., test program) over a reasonably short time (within the ENS reporting time limit for ENS reports, or within the first 30 days of discovery of the first reportable event for LER reporting).

For an outage that lasts longer than 30 days, such as 60 days, similar events that are part of the same activity or test program and are therefore related may be reported as a single LER.

To the extent feasible, report failures that occurred within the first 30 days of discovery of the first failure on one LER. State in the LER text that a supplement to the LER will be submitted when the test is completed. Include all the failures, including those reported in the original LER, in the revised LER (i.e., the revised LER should stand alone).

Generally, LERs are intended to address specific events and plant conditions. Thus, unrelated events or conditions should not be reported in one LER. Also, an LER revision should not be used to report subsequent failures of the same or like components that are the result of a different cause or for separate events or activities.

Unrelated failures or events should be reported as separate ENS notifications to be given unique ENS numbers by the NRC. However, multiple ENS notifications may be addressed in a single telephone call.

2.4   Deficiencies Discovered During Engineering Reviews or Inspections

As indicated in NUREG-1397, "An Assessment of Design Control Practices and Design Reconstitution Programs in the Nuclear Power Industry," February 1991, Section 4.3.2, the reporting requirements specified in 10 CFR 50.9, 50.72, and 50.73 apply equally to discrepancies discovered during design document reconstitution (DDR) programs, design-bases documentation reviews (DBDRs), and other similar engineering reviews. There is no basis for treating discrepancies discovered during such reviews differently from any other reportable item.

Licensees should evaluate the reportability of suspected but unsubstantiated discrepancies discovered during such a review program in the same manner as other potentially reportable items. See Section 2.11 for discussion of reporting time limits and discovery dates.

2.5   Engineered Safety Features Actuations

Systems typically reported under this criterion include the systems listed in Table 2. The NRC staff considers these systems to be a reasonable interpretation of what constitutes systems "provided to mitigate the consequences of a significant event." Further guidance beyond that provided in this revision is being deferred pending rulemaking to address issues such as whether the rule should list specific systems or provide general guidance. See Section 3.3.2 for further discussion of this matter.

2.6   Events Discussed with the NRC Staff

Some licensee personnel have erroneously believed that if a reportable event or condition had been discussed with the resident inspector or other NRC staff, there was no need to report under 10 CFR 50.72 and 50.73 because the NRC was aware of the situation. Some licensee personnel have also expressed a similar understanding for cases in which the NRC staff identified a reportable event or condition to the licensee via inspection or assessment activities. Such means of reporting do not satisfy 10 CFR 50.72 and 50.73. The requirement is to report to the ENS and LER systems events or conditions meeting the criteria stated in the rules.

2.7 Multiple Component Failures

There have been cases in which licensees have not reported multiple, sequentially discovered failures of systems or components occurring during planned testing. This situation was identified as a generic concern on April 13, 1985, in NRC Information Notice (IN) 85-27, "Notifications to the NRC Operations Center and Reporting Events in Licensee Event Reports," regarding the reportability of multiple events in accordance with §§50.72(b)(2)(iii) and 50.73(a)(2)(v) (event or condition that alone could prevent fulfillment of a safety function). (This reporting criterion is discussed in Section 3.3.3 of this report.)

IN 85-27 described multiple failures of a reactor protection system during control rod insertion testing of a reactor at power. One of the control rods stuck. Subsequent testing identified 3 additional rods that would not insert (scram) into the core and 11 control rods that had an initial hesitation before insertion. The licensee considered each failure as a single random failure; thus each was determined not to be reportable. Subsequent assessments indicated that the instrument air system, which was to be oil-free, was contaminated with oil that was causing the scram solenoid valves to fail. While the failure of a single rod to insert may not cause a reasonable doubt that other rods would fail to insert, the failure of more than one rod does cause a reasonable doubt that other rods could be affected, thus affecting the safety function of the rods.

As indicated in IN 85-27, multiple failures of redundant components of a safety system are sufficient reason to expect that the failure mechanism, even though not known, could prevent the fulfillment of the safety function.

2.8 Preparation of Licensee Event Reports (LERs)

This revision includes new guidance to address consistency of information provided in LERs which is used to understand reported events. Some of this guidance was not specifically addressed in the second draft that was published for comment. It is included to help ensure that consistent information is provided regarding human factors analysis and risk-informed regulatory programs such as accident sequence precursor (ASP) analysis and equipment reliability estimates. It does not affect the decision as to whether or not an event must be reported. See Section 5 for further discussion.

2.9 Voluntary Reporting

Information that does not meet the reporting criteria of 10 CFR 50.72 and 50.73 may be reportable under other requirements such as 10 CFR 50.9, 20.2202, 20.2203, 50.36, 72.74, 72.216, 73.71, and Part 21. In particular, 10 CFR 50.9 (b) states "Each applicant or licensee shall notify the Commission of information identified by the applicant or licensee as having for the regulated activity a significant implication for public health and safety or common defense and security." This applies to information which is not already required by other reporting or updating requirements. Notification must be made to the Administrator of the appropriate Regional Office within two working days of identifying the information. Reporting pursuant to

§50.9 is required, not voluntary.[1] Voluntary reporting, as discussed in the following paragraphs, pertains to information of lesser significance than described in §50.9(b).

The Statement of Considerations for 10 CFR 50.73 states "...licensees are permitted and encouraged to report any event or condition that does not meet the criteria contained in §50.73(a), if the licensee believes that the event or condition might be of safety significance or of generic interest or concern. Reporting requirements aside, assurance of safe operation of all plants depends on accurate and complete reporting by each licensee of all events having potential safety significance."[2] Instructions for completing voluntary LERs are discussed in Section 5.1.5 of this report. In addition, voluntary reporting is encouraged under 10 CFR 50.72, as discussed in Section 4.2.2 of this report.

The NRC staff encourages voluntary LERs rather than information letters to report operational events that do not meet the criteria contained in 10 CFR 50.73. The LER format is preferable because it provides for the information needed to support NRC review of the event and facilitates administrative processing, including data entry.

2.10 Retraction/Cancellation of Event Reports

Licensees have expressed concerns about the counting of event reports, both ENS notifications and LERs. The NRC staff has indicated that its interest is in evaluating the reported information, not in simply counting the number of events reported. While event reports may be formally withdrawn, the staff has often found the information reported useful and has maintained the information on file with the withdrawal notation.

If a licensee so chooses, an ENS notification can be retracted via a follow-up ENS call. LER retractions should be made by letter. The retractions and cancellations are further discussed in Section 4 for ENS notifications and Section 5 for LERs. Sound, logical bases for the withdrawal should be communicated with the retraction or cancellation. (Example 3 in Section 3.3.1 illustrates a case where there were sound reasons for a retraction. The last event under Example 1 in Section 3.3.2 illustrates a case where the reasons for retraction were not adequate.)

2.11 Time Limits for Reporting

Reporting times in 10 CFR 50.72 are keyed to the occurrence of the event or condition, as described below.

---

[1]   As indicated in the Statement of Considerations for §50.9, "A licensee cannot evade the rule by never 'finding' information to be significant. The fact that a licensee considers information to be significant can be established, for example, by the actions taken by the licensee to evaluate that information." 59 FR 49362, December 31, 1987.

[2]   48 FR 33853, July 26, 1983.

Section 50.72(a)(3) requires ENS notification of the declaration of an Emergency Class "...immediately after notification of the appropriate State or local agencies and not later than one-hour after the time the licensee declares one of the Emergency Classes."

Section 50.72(b)(1) requires ENS notification for specific types of events and conditions "...as soon as practical and in all cases, within one-hour of the occurrence of any of the following:...."

Section 50.72(b)(2) requires ENS notification for specific types of events and conditions "...as soon as practical and in all cases, within four hours of the occurrence of any of the following:...."

Section 50.73 requires submittal of an LER "within 30 days after the discovery" of a reportable event.

Many reportable events are discovered when they occur. However, if the event is discovered at some later time, the discovery date is when the reportability clock starts under 10 CFR 50.73.

Discovery date is generally the date when the event was discovered rather than the date when an evaluation of the event is completed. For example, if a technician sees a problem, but a delay occurs before an engineer or supervisor has a chance to review the situation, the discovery date (which starts the 30-day clock) is the date that the technician sees a problem.

In some cases, such as discovery of an existing but previously unrecognized condition, it may be necessary to undertake an evaluation in order to determine if an event or condition is reportable. If so, the guidance provided in Generic Letter 91-18, "Information to Licensees Regarding two NRC Inspection Manual Sections on Resolution of Degraded and Nonconforming Conditions and on Operability," which applies primarily to operability determinations, is appropriate for reportability determinations as well. This guidance indicates that, whenever reasonable expectation that the equipment in question is operable no longer exists, or significant doubts begin to arise, appropriate actions, including reporting, should be taken.

2.12 Outside Design Basis

This revision provides new guidance and several examples for this reporting criterion and two related criteria, seriously degraded condition and unanalyzed condition that seriously compromised plant safety. Further guidance beyond that provided in this revision is being deferred pending rulemaking to address issues such as: (1) one-hour reporting for design basis issues, (2) significance testing for reporting design basis issues, and (3) scope of plant design basis. See Section 3.2.4 of this report for further discussion.

3 SPECIFIC REPORTING GUIDELINES

This section addresses the specific requirements of each part of the rules cited for immediate notification of an event under 10 CFR 50.72 via the ENS and 30-day written reports under 10 CFR 50.73 via LERs. The section is divided into four parts. Section 3.1 gives the general requirements for reporting, Section 3.2 gives the criteria for 1-hour notifications and 30-day reports, Section 3.3 gives the criteria for 4-hour notifications and 30-day reports, and Section 3.4 addresses followup notifications.

The sequential scheme of 10 CFR 50.72 is used, which generally categorizes the times for reporting by the relative importance of the event or condition. Because considerable overlap exists between the various reporting criteria in each rule, the associated requirements for licensee event reporting (10 CFR 50.73) are given coincidentally. Differences in the wording of the comparable parts of the rules are underlined. In several instances, the wording of the two rules is the same except for verb tense. A discussion of reporting guidelines and examples follow each citation of specific parts of the rules. Brief examples occasionally are given in the discussion for clarification; however, expanded examples for each part of the rules are discussed under "Examples." The descriptions in the expanded examples have been taken from actual operational experience and have been condensed to illustrate specific aspects of reportability.

The reporting requirements in each of the two rules are not mutually exclusive, and many events and conditions are reportable under more than one criterion. Therefore, it is important to first recognize whether an event or condition is reportable under at least one criterion, and then to identify other applicable criteria. When the report is made to the NRC, applicable criteria should be cited.

3.1 Section 50.72 and 50.73 General Requirements

3.1.1 Immediate Notification Requirements

§50.72(a) General Requirements[1]	10 CFR 50.73

"(1) Each nuclear power reactor licensee licensed under §50.21(b) or §50.22 of this part shall notify the NRC Operations Center via the Emergency Notification System of:
    (i) The declaration of any of the Emergency Classes specified in the licensee's approved Emergency Plan;[2] or
    (ii) Of those non-Emergency events specified in paragraph (b) of this section.
    (2) If the Emergency Notification System is inoperative, the licensee shall make the required notifications via commercial telephone service, other dedicated telephone system, or any other method which will ensure that a report is made as soon as practical to the NRC Operations Center.[3]
    (3) The licensee shall notify the NRC immediately after notification of the appropriate State or local agencies and not later than one hour after the time the licensee declares one of the Emergency Classes.
    (4) The licensee shall activate the Emergency Response Data System (ERDS)[5] as soon as possible but not later than one hour after declaring an emergency class of alert, site area emergency, or general emergency. The ERDS may also be activated by the licensee during emergency drills or exercises if the licensee's computer system has capability to transmit the exercise data."
	[If the event or condition that was the basis for the Emergency Class declaration met one or more of the 10 CFR 50.73 reporting criteria, an LER is required.]

_____

"[1] Other requirements for immediate notification of the NRC by licensed operating nuclear power reactors are contained elsewhere in this chapter, in particular, §§ 20.1906, 20.2202, 50.36, and 73.71.

[2] These Emergency Classes are addressed in Appendix E of this part.

[3] Commercial telephone number of the NRC Operations Center is (301) 816-5100."

[4] [Reserved]

[5] Requirements for ERDS are addressed in Appendix E, Section VI."

(Continued on next page)

<table>
<tr><td>

50.72(a) (Continued)

"(5) When making a report under paragraph (a)(3) of this section, the licensee shall identify
    (i) The Emergency Class declared; or
    (ii) Either paragraph (b)(1), "One-Hour Report," or paragraph (b)(2), "Four-Hour Report," as the paragraph of this section requiring notification of the Non-Emergency Event."

</td><td>

</td></tr>
</table>

Discussion

Appendix E to 10 CFR Part 50, Section IV (C), "Activation of Emergency Organization," establishes four emergency classes for nuclear power plants: Notification of Unusual Event, Alert, Site Area Emergency, and General Emergency. NUREG-0654/FEMA-REP-1, Revision 1, "Criteria for Preparation and Evaluation of Radiological Emergency Response Plans and Preparedness in Support of Nuclear Power Plants" (March 1987), and more recently, NUMARC/NESP-007, Revision 2, "Methodology for Development of Emergency Action Levels" (January 1992), provides the basis for these emergency classes and numerous examples of the events and conditions typical of each emergency class. Licensees use this guidance in preparing their emergency plans. Use of these four emergency class terms in the ENS notification will help the NRC recognize the significance of an emergency. Time frames specified for notification in §50.72(a) use the words "immediately" and "not later than one hour" to ensure the Commission can fulfill its responsibilities during and following the most serious events.

Occasionally, a licensee may discover that an event or condition had existed which met the emergency plan criteria but that no emergency had been declared and the basis for the emergency class no longer exists at the time of this discovery. This may be due to a rapidly concluded event or an oversight in the emergency classification made during the event or it may be determined during a post-event review. Frequently, in cases of this nature, which were discovered after the fact, licensees have declared the emergency class, immediately terminated the emergency class and then made the appropriate notifications. However, the staff does not consider actual declaration of the emergency class to be necessary in these circumstances; an ENS notification (or an ENS update if the event was previously reported but misclassified) within one hour of the discovery of the undeclared (or misclassified) event will provide an acceptable alternative.[3]

---

[3]   Notification of the State and local emergency response organizations should be made in accordance with the arrangements made between the licensee and offsite organizations.

3.1.2 Licensee Event Report System

10 CFR 50.72	§50.73(a)(1)
[Bases for ENS notifications (e.g., regardless of plant status), are the same as 10 CFR 50.73 where the two rules are complementary.]	"The holder of an operating license for a nuclear power plant (licensee) shall submit a Licensee Event Report (LER) for any event of the type described in this paragraph within 30 days after the discovery of the event. Unless otherwise specified in this section, the licensee shall report an event regardless of the plant mode or power level, and regardless of the significance of the structure, system, or component that initiated the event."

Discussion

Unless otherwise specified, this part of the rule requires reporting of an event <u>regardless</u> of the plant mode or power level and <u>regardless</u> of the significance of the structure, system, or component that initiated the event. These considerations also are implicit in 10 CFR 50.72 where the two rules are complementary.

3.2 One-hour ENS Notifications and 30-Day LERs

This section addresses §50.72(b)(1) 1-hour notifications for non-emergency events and the associated 10 CFR 50.73 written reports. If not reported as a declaration of an emergency class under §50.72(a), licensees are to notify the NRC as soon as practical and in all cases within 1 hour of the occurrence of any of the events specified in §50.72(b)(1) and to submit an LER, if specified.

In addition to similar reporting criteria under both 10 CFR 50.72 and 50.73, several requirements for only 50.72 notifications or only LERs are included in this section because of the sequential numbering scheme used. For example, operation or a condition prohibited by the plant's technical specifications (TS), as discussed in Section 3.2.2, requires an LER but no ENS notification, while loss of emergency assessment, response or communications capability, as discussed in Section 3.2.7, requires an ENS notification but no LER.

3.2.1 Plant Shutdown Required by Technical Specifications

§50.72(b)(1)(i)(A)	§50.73(a)(2)(i)(A)
Licensees shall report: "The initiation of any nuclear plant shutdown required by the plant's Technical Specifications."	Licensees shall submit a Licensee Event Report on: "The completion of any nuclear plant shutdown required by the plant's Technical Specifications."

If not reported as an emergency under §50.72(a), licensees are required to report the initiation of a plant shutdown required by TS to the NRC via the ENS as soon as practical and in all cases within 1 hour of the initiation of a plant shutdown required by TS to the NRC via the ENS. If the shutdown is completed, licensees are required to submit an LER within 30 days.

Discussion

The §50.72 reporting requirement is intended to capture those events for which TS require the initiation of reactor shutdown to provide the NRC with early warning of safety significant conditions serious enough to warrant that the plant be shut down.

For §50.72 reporting purposes, the phrase "initiation of any nuclear plant shutdown" includes action to start reducing reactor power, i.e., adding negative reactivity to achieve a nuclear plant shutdown required by TS. The "initiation of any nuclear plant shutdown" does not include mode changes required by TS if initiated after the plant is already in a shutdown condition.

A reduction in power for some other purpose, not constituting initiation of a shutdown required by TS, is not reportable under this criterion.

For §50.73 reporting purposes, the phrase "completion of any nuclear plant shutdown" is defined as the point in time during a TS required shutdown when the plant enters the first shutdown condition required by a limiting condition for operations (LCO) e.g., hot standby {Mode 3] for PWRs with the standard technical specifications (STS). For example, if at 0200 hours a plant enters an LCO action statement that states, "restore the inoperable channel to operable status within 12 hours or be in at least Hot Standby within the next 6 hours," the plant must be shut down (i.e., at least in hot standby) by 2000 hours. An LER is required if the inoperable channel is not returned to operable status by 2000 hours and the plant enters hot standby.

An LER is not required if a failure was or could have been corrected before a plant has completed shutdown (as discussed above) and no other criteria in §50.73 apply.

Examples

(1)   Initiation of a TS-Required Plant Shutdown

While operating at 100-percent power, one of the battery chargers, which feeds a 125 Vdc vital bus, failed during a surveillance test. The battery charger was declared inoperable, placing the plant in a 2-hour LCO to return the battery charger to an operable status or commence a TS-required plant shutdown. Licensee personnel started reducing reactor power to achieve a nuclear plant shutdown required by a TS when they were unable to complete repairs to the inoperable battery charger in the 2 hours allowed. The cause of the battery charger failure was subsequently identified and repaired. Upon completion of surveillance testing, the battery charger was returned to service and the TS required plant shutdown was stopped at 96-percent power.

The licensee made an ENS notification because of the initiation of a TS-required plant shutdown An LER was not submitted under this criterion since the failed battery charger was corrected before the plant completed shutdown.

(2)   Initiation and Completion of a TS-Required Plant Shutdown

During startup of a PWR plant with reactor power in the intermediate range, two of the four reactor coolant pumps (RCPs) tripped when the station power transformer supplying power, deenergized. With less than four RCPs operating, the plant entered a 1-hour LCO to be in hot standby. Control rods were manually inserted to place the plant in a shutdown condition.

The licensee made an ENS notification because of the initiation of a TS-required plant shutdown. An LER was submitted within 30 days because of the completion of the TS-required plant shutdown.

(3)   Failure that was or could have been corrected before a plant has completed shut down.

*   Question:

What about the situation where you have seven days to fix a component or be shut down, but the plant must be shut down to fix the component? Assume the plant shuts down, the component is fixed, and the plant returns to power prior to the end of the seven day period. Is that situation reportable?

Answer:

No. If the shutdown was not required by the Technical Specifications, it need not be reported. However, other criteria in 50.73 may apply and may require that the event be reported.

●   Question:

Suppose that there are seven days to fix a problem and it is likely the problem can be fixed during this time period. However, the plant management elects to shut down and fix this problem and other problems. It an LER required?

Answer:

Some judgment is required. An LER is not required if the situation could have been corrected before the plant was required to be shut down, and no other criteria in 50.73 apply. The shut down is reportable, however, if the situation could not have been corrected before the plant was required to be shut down, or if other criteria of 50.73 apply.

3.2.2 Technical Specification Prohibited Operation or Condition

10 CFR 50.72	§50.73(a)(2)(i)(B)
[There is no corresponding Part 50.72 requirement. However, for certain operations or conditions prohibited by a plant's TS, other reporting requirements may apply, such as 50.72(b)(1)(ii) and (b)(2)(iii); 50.36(c)(1) and (2); 20.2202; and 20.2203.]	Licensees shall report: "any operation or condition prohibited by the plant's Technical Specifications."

Licensees are required to submit an LER within 30 days for any operation or condition prohibited by technical specifications.

Discussion

Section 50.73(a)(2)(i)(B) requires any operation or condition that is prohibited by the plant's TS to be reported in an LER. The five specific TS categories defined in 10 CFR 50.36(c), "Technical Specifications," are discussed below. In addition, based on past experience, guidelines are provided for reporting entry into TS 3.0.3 [ISTS [4] Limiting Condition for Operation (LCO) 3.0.3]; missed or deficient tests required by the American Society of Mechanical Engineers (ASME) Section XI, Inservice Testing (IST) and Inservice Inspection (ISI), and by STS 4.0.5, or equivalent; and other operations or conditions prohibited by TS, such as fire protection.

The LER rule does not address violations of license conditions contained in documents other than the TS. Such notifications are reportable as specified in a plant's license or other applicable documents.

(1)   Safety Limits and Limiting Safety System Settings

Section 50.36(c)(1) outlines the reporting requirements in TS when nuclear reactor safety limits or limiting safety system settings are exceeded and identifies that such reports are to be made under 50.72 and 50.73.

---

[4]   To be consistent with the improved Standard Technical Specifications (ISTS) discussed in the NUREG-1430 through NUREG-1434 (e.g., NUREG-1431, Vol. 1, Standard Technical Specifications - Westinghouse Plants, September 1992) references to appropriate sections in these ISTS have been included throughout this section of NUREG-1022. The designation used here for references to such sections is "ISTS" followed by the appropriate section number.

(2)   Limiting Conditions for Operation

Section 50.36(c)(2) outlines LCOs in TS. Certain TS contain LCO statements that include action statements (required actions and associated completion time in ISTS) to provide constraints on the length of time components or systems may remain inoperable or out of service before the plant must shut down or other compensatory measures must be taken. Such time constraints are based on the safety significance of the component or system being removed from service.

An LER is required if the conditions of an LCO are not met, e.g., by exceeding action statement constraints (not meeting required actions and associated completion times in ISTS).

The LCO allows a plant a specific time interval referred to as the allowed outage time (or completion time in ISTS) to accomplish corrective actions (e.g., restoration of equipment, testing of other equipment, and/or an orderly shutdown to either the hot- or cold-shutdown mode or operating condition).

If a condition existed for a time longer than permitted by the TS [i.e., greater than the allowed outage time (or completion time in ISTS)] it must be reported even if the condition was not discovered until after the allowable time had elapsed and the condition was rectified immediately upon discovery. This guidance is consistent with that previously given. (For the purpose of this discussion, it is assumed that there was firm evidence that a condition prohibited by TS existed before discovery, for a time longer than permitted by TS.)

(3)   TS Surveillance Requirements

Section 50.36(c)(3) outlines surveillance requirements in TS which assure (1) necessary quality of systems and components, (2) operation within safety limits, and (3) that the limiting conditions for operation will be met. For the purpose of evaluating the reportability of discrepancies found during TS surveillance tests, an operation or condition prohibited by the TS existed and is reportable if the time of equipment inoperability exceeded the LCO allowed outage time (or completion time for restoration of equipment in ISTS). It should be assumed that the discrepancy occurred at the time of its discovery unless there is firm evidence, based on a review of relevant information (e.g., the equipment history and cause of failure) to believe that the discrepancy existed previously. As discussed in Example 5, evaluation of multiple similar failures may indicate that a condition has persisted for some time.

Missed surveillance tests are reportable when the surveillance interval plus allowed surveillance interval extension, e.g., STS section 4.0.2 (or ISTS SR 3.0.2), plus the LCO action statement time is exceeded.[5] This means that a condition prohibited by TS

---

[5]   The Statement of Considerations for the final rule (48 FR 33855, July 28, 1983, Second column) states, in part, ". . . if a condition that was prohibited by the Technical

existed for a period of time longer than allowed by TS. The event is reportable even though the surveillance is subsequently satisfactorily performed.[6]

Some plants have TS which allow a delay of up to 24 hours in declaring an LCO or TS requirements not met if it is found that a surveillance was not performed within its specified frequency or interval. However, failure to perform a surveillance within its frequency or interval is still reportable. The additional delay in declaring the LCO not met does not change the fact that the condition existed longer than allowed by TS. The delay merely specifies appropriate remedial action.

(4)   Design Features

Section 50.36(c)(4) indicates that design features to be included in TS are those features of the facility such as materials of construction or geometric arrangements which, if altered or modified, would have a significant effect on safety and are not covered by items (1) through (3) above.

Reportability requirements related to design features are included in other sections of 10 CFR 50.72 and 50.73.

(5)   TS Administrative Requirements, Including Radiological Controls, Required by Section 6 of the STS, or Equivalent

Section 6 of the STS (Section 5 of ISTS), or its equivalent, has a number of administrative requirements such as organizational structure, the required number of personnel on shift, the maximum hours of work permitted during a specific interval of time, and the requirement to have, maintain, and implement certain specified procedures.

Failure to meet such administrative requirements is prohibited by the TS. Whether it is reportable as an LER depends upon whether it results in a condition covered by the LER rule. If the violation of the administrative requirements of TS results in operations prohibited by TS, then it is reportable.

Violation of an administrative TS in and of itself does not necessarily constitute a reportable condition ("operation or condition prohibited by the plant's TS"). This

---

Specifications existed for a period of time longer than that permitted by the Technical Specifications, it must be reported even if the condition was not discovered until after the allowable time had elapsed and the condition was rectified immediately after discovery."

[6]   This guidance is only intended to define when the matter becomes reportable under this specific reporting criterion ("operation or condition prohibited by the plant's TS"). It is not intended to define when a TS violation occurs, when a system must actually be declared inoperable, when the surveillance must be completed, or when the plant must be shutdown. These matters are discussed further in GL 87-09, GL 91-18, TS 4.0.2, and ISTS SR 3.0.3.

reporting requirement deals with matters affecting plant operation more substantially and more directly than matters that are mainly administrative.[7] Failure to meet administrative TS requirements is reportable only if it results in violations of equipment operability requirements, or had a similar detrimental effect on a licensee's ability to safely operate the plant.

For example, operation with less than the required number of people on shift would constitute operation prohibited by the TS. However, a change in the plant's organizational structure that has not yet been approved as a Technical Specification change would not.

An administrative procedure violation or failure to implement a procedure, such as failure to lock a high radiation area door, that does not have a direct impact on the safe operation of the plant, is generally not reportable under this criterion.

Radiological conditions and events that are reportable are defined in 10 CFR 20.2202 and 20.2203. Redundant reporting is not required.

(6)   Entry into STS 3.0.3

STS 3.0.3 (ISTS LCO 3.0.3), or its equivalent, establishes requirements for actions when an LCO is not met and no action statement is provided. Entry into STS 3.0.3 is considered to be the action taken, as required, when operations or conditions required by TS LCO action statements (ISTS required actions and associated completion times) are not met. Thus, entry into STS 3.0.3 (ISTS LCO 3.0.3) for any reason or justification is reportable.

(7)   Missed Tests Required by ASME Section XI IST and ISI and by STS 4.0.5, or Equivalent

Sections 50.55a(g) and 50.55a(f) require the implementation of ISI and IST programs in accordance with the applicable edition of the ASME Code for those pumps and valves whose function is required for safety. STS Section 4.0.5 (or an equivalent) covers these testing requirements. (Generally, there is no comparable ISTS section.) Missed IST/ISI/ASME tests are reportable when the test interval plus any allowable extension plus the LCO action time has been exceeded.

---

[7]   The proposed rule would have required reporting when "a TS action statement is not met." The wording of the final rule requires reporting "Any operation or condition prohibited by the plant's Technical Specifications." The Statement of Considerations for the final rule indicates that this change was made to accommodate plants that did not have requirements specifically defined as action statements (48 FR 33855, July 26, 1983).

(8)   Fire Protection Systems When Required by TS

        When operability requirements for fire protection systems are specified in TS they are
        within the scope of this reporting criterion.

Examples

(1)   LCO Exceeded

        A licensee found a standby component with a 7-day LCO allowed outage time and
        associated 8-hour shutdown action statement to be inoperable during a 30-day
        surveillance test. (This is equivalent to a 7-day restoration completion time and an 8-
        hour action completion time in ISTS.) Subsequent review indicated that the component
        was assembled improperly during maintenance conducted 30 days previously and the
        post-maintenance test was not adequate to identify the error. Thus, there was firm
        evidence that the standby component had been inoperable for the entire 30 days.

        An LER was required because the 7-day LCO allowed outage time and the shutdown
        action statement time of 8 hours were exceeded. Had the inoperability been identified
        and corrected within the 7-day LCO allowed outage time plus the 8-hour shutdown
        action statement, the event would not be reportable.

(2)   Missed Surveillance Tests

        A licensee, with the plant in Mode 5 following a 10-month refueling outage, determined
        that certain monthly TS surveillance tests, which were required to be performed
        regardless of plant mode, had not been performed as required during the outage. The
        STS 4.0.2 (equivalent to ISTS SR 3.0.2) extension was also exceeded. The
        surveillance tests were immediately performed. An LER is required because the time
        interval, including extensions permitted by TS, exceeded the TS surveillance interval
        plus the LCO action statement times (equivalent to ISTS completion times).

(3)   Entering STS 3.0.3

        With essential water chillers (A) and (B) out of service, the only remaining operable
        chiller (A/B) tripped. This condition caused the plant to enter STS 3.0.3 (equivalent to
        ISTS LCO 3.0.3) for 1 hour until chiller (A) was restored to service and the temperature
        was restored to within TS limits. An LER is required for this event because STS 3.0.3
        was entered.

(4)   Missed Tests Required by ASME Section XI IST and ISI, and by STS 4.0.5, or
        Equivalent

        Examples of potentially reportable conditions are failures to perform required activities
        within specified times for those components governed by TS. Such activities include
        stroke testing valves, testing valves in the position required for the performance of their
        safety function, verifying motor-operated valve stroke times for both (open and closed)

directions, using the proper test pressures to properly classify and test active valves and to increase test frequency subsequent to obtaining test results that were below certain threshold values. A missed test is reportable when the test interval plus any allowable extension plus the LCO action time is exceeded.

(5)   Multiple Test Failures

An example of multiple test failures involves the sequential testing of safety valves. Sometimes multiple valves are found to lift with setpoints outside of TS limits.

As discussed above, discrepancies found in TS surveillance tests should be assumed to occur at the time of the test unless there is firm evidence, based on a review of relevant information (e.g., the equipment history and the cause of failure) to believe that the discrepancy occurred earlier. However, the existence of similar discrepancies in multiple valves is an indication that the discrepancies arose over a period of time. Therefore, the condition existed during plant operation and the event is reportable under §50.73(a)(2)(i)(B) "Any operation or condition prohibited by the plant's Technical Specifications."

If the discrepancies are large enough that multiple valves are inoperable the event may also be reportable under §50.73(a)(2)(vii) "Any event where a single cause or condition caused at least one independent train or channel to become inoperable in multiple systems or two independent trains or channels to become inoperable in a single system ...."

3.2.3 Technical Specification Deviation per §50.54(x)

§50.72(b)(1)(i)(B)	§50.73(a)(2)(i)(C)
Licensees shall report: "Any deviation from the plant's Technical Specifications authorized pursuant to §50.54(x) of this part."	Licensees shall report: "Any deviation from the plant's Technical Specifications authorized pursuant to §50.54(x) of this part."

If not reported as an emergency under §50.72(a), licensees are required to report any such deviation to the NRC via the ENS as soon as practical and in all cases within 1 hour. Licensees are required to submit an LER within 30 days.

Discussion

10 CFR 50.54(x) generally permits licensees to take reasonable action in an emergency even though the action departs from the license conditions or plant technical specifications if (1) the action is immediately needed to protect the public health and safety, including plant personnel, and (2) no action consistent with the license conditions and technical specifications is immediately apparent that can provide adequate or equivalent protection. Deviations authorized pursuant to 10 CFR 50.54(x) are reportable under this criterion.

Example

With the plant at 100-percent power, the upper containment airlock inner door was opened to allow a technician to exit from the containment while the upper airlock outer door was inoperable, resulting in the loss of containment integrity. The upper airlock door was inoperable pending retests following seal replacement. The technician was inside containment when the lower airlock failed, requiring the technician to exit through the upper door.

The licensee decided to exercise the option allowed for under 10 CFR 50.54(x) and open the upper containment airlock inner door. In this instance, immediate action was considered necessary to protect the safety of the technician. The upper airlock was not scheduled to be returned to operability for another 20 hours and the time to repair the lower airlock door was unknown. When the action was completed the control room operators notified the NRC Operations Center, in accordance with the reporting requirements of 10 CFR 50.72, that they had exercised 10 CFR 50.54(x).

Subsequently, an LER was submitted in accordance with 10 CFR 50.73(a)(2)(i) {use of 10 CFR 50.54(x)} as well as 10 CFR 50.73(a)(2)(v) {event or condition that alone could prevent ....}.

3.2.4 Operating Plant Found in Degraded or Unanalyzed Condition

§50.72(b)(1)(ii)	§50.73(a)(2)(ii)

Licensees shall report: "Any event or condition <u>during operation</u> that result<u>s</u> in the condition of the nuclear power plant, including its principal safety barriers, being seriously degraded; or result<u>s</u> in the nuclear power plant being:
  (A) In an unanalyzed condition that significantly compromise<u>s</u> plant safety;
  (B) In a condition that <u>is</u> outside the design basis of the plant; or
  (C) In a condition not covered by the plant's operating and emergency procedures."
	Licensees shall report: "Any event or condition that result<u>ed</u> in the condition of the nuclear power plant, including its principal safety barriers, being seriously degraded; or <u>that</u> result<u>ed</u> in the nuclear power plant being:
	  (A) In an unanalyzed condition that significantly compromise<u>d</u> plant safety;
	  (B) In a condition that <u>was</u> outside the design basis of the plant; or
	  (C) In a condition not covered by the plant's operating and emergency procedures."

If not reported as an emergency under §50.72(a), licensees are required to report operation under such a condition to the NRC via the ENS as soon as practical and in all cases within 1 hour. Licensees are required to submit an LER within 30 days.

Discussion

Reporting at the component, system, and structure level is required under 10 CFR 50.72(b)(1)(ii) and 50.73(a)(2)(ii) if the event or condition resulted in the plant being seriously degraded, in an unanalyzed condition that significantly compromises plant safety, outside the plant design basis, or in a condition not covered by the plant's procedures, as described in the rule.

The discussions below provide further guidance on reportability under these criteria.

(1)   *The condition of the nuclear power plant, including its principal safety barriers, being seriously degraded.*

    As indicated in the Statements of Considerations, this paragraph includes material (e.g., metallurgical or chemical) problems that cause abnormal degradation of the principal safety barriers (i.e., the fuel cladding, reactor coolant system pressure boundary, or the containment). Examples of this type of situation include:

    (a)   Fuel cladding failures in the reactor, or in the storage pool, that exceed expected values, or that are unique or widespread, or that are caused by unexpected factors, and would involve a release of significant quantities of fission products.

(b)   Cracks and breaks in the piping or reactor vessel (steel or prestressed concrete) or major components in the primary coolant circuit that have safety relevance (steam generators, reactor coolant pumps, valves, etc).

(c)   Significant welding or material defects in the primary coolant system.

(d)   Serious temperature or pressure transients.

(e)   Loss of relief and/or safety valve functions during operation.

(f)   Loss of containment function or integrity including:

        (i)   Containment leakage rates exceeding the authorized limits.

        (ii)   Loss of containment isolation valve function during tests or operation.

        (iii)   Loss of main steam isolation valve function during test or operation, or

        (iv)   Loss of containment cooling capability.

Examples of events that the staff would consider reportable as significant reactor coolant system welding or material defects include items which cannot be found acceptable under ASME Section XI, IWB-3600, "Analytical Evaluation of Flaws" or ASME Section XI, Table IWB-3410-1, "Acceptance Standards."

Examples of events that the staff would consider reportable as serious temperature or pressure transients include low temperature over pressure transients where the pressure-temperature relationship violates pressure-temperature limits derived from Appendix G to 10 CFR Part 50 (e.g., TS pressure-temperature curves).

Examples of events the staff would consider reportable as containment leakage rates exceeding authorized limits include containment leak rate tests where the total containment as-found, minimum-pathway leak rate exceeds the LCO in the facility's TS.[8],[9]

---

[8]   The LCO typically employs La, which is defined in Appendix J to 10 CFR Part 50 as the maximum allowable containment leak rate at pressure Pa, the calculated peak containment internal pressure related to the design basis accident. Minimum-pathway leak rate means the minimum leak rate that can be attributed to a penetration leakage path; for example, the smaller of either the inboard or outboard valve's individual leak rates.

[9]   For such a condition, an LER is generally required under 10 CFR 50.73(a)(2)(ii). If the condition existed during operation, an ENS notification would also be required by §50.72(b)(1)(ii) if found during operation or by §50.72(b)(2)(i) if found while shutdown.

(2)   *The nuclear power plant being in an unanalyzed condition that significantly compromises plant safety.*

As indicated in the Statements of Consideration:

"The Commission recognizes that the licensee may use engineering judgment and experience to determine whether an unanalyzed condition existed. It is not intended that this paragraph apply to minor variations in individual parameters, or to problems concerning single pieces of equipment. For example, at any time, one or more safety-related components may be out of service due to testing, maintenance, or a fault that has not yet been repaired. Any trivial single failure or minor error in performing surveillance tests could produce a situation in which two or more often unrelated, safety-grade components are out-of-service. Technically, this is an unanalyzed condition. However, these events should be reported only if they involve functionally related components or if they significantly compromise plant safety."[10]

"When applying engineering judgment, and there is a doubt regarding whether to report or not, the Commission's policy is that licensees should make the report."[11]

"For example, small voids in systems designed to remove heat from the reactor core which have been previously shown through analysis not to be safety significant need not be reported. However, the accumulation of voids that could inhibit the ability to adequately remove heat from the reactor core, particularly under natural circulation conditions, would constitute an unanalyzed condition and would be reportable."[12]

"In addition, voiding in instrument lines that results in an erroneous indication causing the operator to misunderstand the true condition of the plant is also an unanalyzed condition and should be reported."[13]

(3)   *The nuclear power plant being in a condition that is outside the design basis of the plant.*

As indicated in 10 CFR 50.2, "Design bases means that information which identifies the specific functions to be performed by a structure, system, or component of a facility, and the specific values or ranges of values chosen for controlling parameters as reference bounds for design. These values may be (1) restraints derived from generally accepted 'state of the art' practices for achieving functional goals, or (2) requirements derived

---

[10]   48 FR 39042, August 29, 1983 and 48 FR 33856, July 26, 1983.

[11]   48 FR 39042, August 29, 1983.

[12]   48 FR 39042, August 29, 1983 and 48 FR 33856, July 26, 1983.

[13]   48 FR 39042, August 29, 1983 and 48 FR 33856, July 26, 1983.

from analysis (based on calculation and/or experiments) of the effects of a postulated accident for which a structure, system, or component must meet its functional goals." (Emphasis added.)

Examples of events or conditions the staff considers reportable include errors in the actual design, such as discovery that an ECCS design does not meet the single failure criterion. They also include hardware problems such as discovery that high energy line break restraints are not installed.

With regard to ECCS calculations, detailed reporting criteria are given in 10 CFR 50.46(a). That rule provides that a change or error correction that results in calculated ECCS performance that does not meet the acceptance criteria {peak cladding temperature, cladding oxidation, etc.} is a reportable event as described in 10 CFR 50.55(e), 50.72 and 50.73. Lesser changes or error corrections are the subject of other, separate reports.

Violation of fire protection commitments regarding safe shutdown capability may indicate that the plant is outside of its design basis. For example, if fire barriers are found to be missing, such that the required degree of separation for redundant safe shutdown trains is lacking, the plant would be outside of the design basis with respect to Appendix R to 10 CFR Part 50. On the other hand, if a fire wrap, to which the licensee has committed, is missing from a safe shutdown train but another safe shutdown train is available in a different fire area, protected such that the required separation for safe shutdown trains is still provided, the plant would not be outside of its design basis with respect to Appendix R.[14]

Another example of an event or condition that the staff considers reportable is discovery that one train of a required two train safety system has been incapable of performing its design function (intended safety function) for an extended period of time during operation. For example, in a two-train ECCS system, one train might be found with a design flaw or with a component that would never have functioned because it was installed incorrectly and a test that would reveal the problem was not performed, such that the train was incapable of performing its design function. This would be considered outside the design basis because the system did not have suitable redundancy.[15]

---

[14]   The design basis with respect to Appendix R for protection of safe shutdown capability is essentially the same as the required protection features. This is discussed in the Statement of Considerations for Appendix R, Federal Register, November 19, 1980 (45 FR 76606). In particular, it is stated that "Because it is not possible to predict the specific conditions under which fires may occur and propagate, the design basis protective features are specified rather than the design basis fire."

[15]   A minimum design feature is suitable redundancy meeting the single-failure criterion as indicated in: (1) 10 CFR Part 50, Appendix A, Introduction and 10 CFR 50, Appendix A, Criterion 35; (2) 10 CFR 50, Appendix K, Item I.D.1; AND (3) FSAR commitments.

It should be noted that these discussions concern events or conditions that actually place the plant outside its design basis. They are not intended to capture minor problems such as: (1) cases of administrative inoperability, where a component is declared inoperable because a surveillance test is overdue but the equipment is actually capable of performing its design function, or (2) cases where the LCO allowed outage time is exceeded by a modest amount (e.g., less than 25 percent). Such conditions may, however, be reportable as conditions prohibited by the Technical Specifications, 10 CFR 50.73(a)(2)(i)(B).

(4) *The nuclear power plant being in a condition not covered by the plant's operating and emergency procedures.*

This criterion points to events where the plant is in a condition outside the coverage of its operating and emergency procedures. A straightforward example of this type of event was the accident at Three Mile Island.

Examples

(1) Design Problem (ECCS Single Failure Vulnerability)

A minimum design feature for ECCS is suitable redundancy meeting the single-failure criterion. Sources include: (1) 10 CFR Part 50, Appendix A, Introduction and 10 CFR 50, Appendix A, Criterion 35; (2) 10 CFR 50, Appendix K, Item I.D.1; and (3) FSAR commitments. During an engineering review following an event, it was found that a coil shorting in one of several supervisory relays, in conjunction with an accident, could lead to a premature recirculation actuation signal. This could result in loss of water to the ECCS pumps due to realignment of suction to the containment sump. It was also found that such coil shorting could cause closure of pump recirculation isolation valves, potentially dead heading and possibly damaging the HPSI pumps.

The licensee determined that each of these concerns represented a condition outside of the design bases. When each determination was made, an ENS notification was made and immediate actions were taken to fix the single failure problem. Subsequently, an LER was submitted.

The event is reportable because the ECCS failed to meet its design bases.

(2) Design/Hardware Problem (Turbine Missile Protection)

The original design criteria, as stated in the UFSAR, required that ESFs be protected from turbine generated missiles by means of shielding or separation. As a result of a service water upgrade project it was found that portions of the low pressure service water system (LPSW) did not meet the plant's separation criteria for high trajectory turbine missiles. The licensee provided an ENS notification under 10 CFR 50.72(b)(1)(ii)(B) and submitted an LER under 10 CFR 50.73(a)(2)(ii)(B), outside design basis.

The corrective action included submitting a UFSAR amendment, which was approved by the NRC staff, to allow using current NRC and industry guidance. When applying this guidance, the LPSW piping in question provides an acceptably low probability target.

This event is reportable because the turbine missile protection did not meet the design basis as stated in the FSAR.

(3)   ECCS Analysis

The large break LOCA analysis, as documented in the FSAR, assumed that high and low head safety injection systems can deliver full flow in 5 and 10 seconds, respectively. A new analysis was performed, which accounted conservatively for (1) SI signal processing, (2) sequencer delay time uncertainty, and (3) increased time for pump acceleration to full speed due to degraded voltage. This indicated that it could take as much as 8 and 24 seconds, respectively, for the high and low head safety injection systems to deliver full flow.

As a result of the new analysis, calculated peak clad temperature was increased by about 44F. However, peak clad temperature remained below 2200F and other ECCS acceptance criteria continued to be met as well. Although licensee reported the event as outside design bases, staff does not consider the event reportable under that criterion because the provisions of §50.46(a) apply. Under those provisions, the events reportable pursuant to §50.72 and §50.73 are those where the ECCS acceptance criteria are exceeded.

The event is not reportable because the provisions of §50.46(a) apply and the ECCS acceptance criteria were not exceeded.

(4)   Fire Protection (Separation of Safe Shutdown Trains)

The design for a fire area had been approved on the basis of several specific features including: automatic sprinklers; remote annunciation; and circuit separation via 2-hour rated fire barriers for redundant safe shutdown circuits. During a design basis review it was found that redundant diesel generator field circuits, located in a common fire area, were not protected or separated by 2-hour rated fire barriers.

An ENS notification was made under 10 CFR 50.72(b)(1)(ii)(B), outside design basis, and subsequently an LER was submitted under 10 CFR 50.73(a)(2)(ii)(B), outside design basis.

The condition was reportable under this criterion because the required design basis protective features for safe shutdown trains, as described in 10 CFR 50, Appendix R and the FSAR, were lacking.

(5)   Hardware Problem (Suitable Redundancy and Seismic Qualification)

During an NRC evaluation, it was found that an exciter panel for one diesel generator had lacked appropriate seismic restraints since the plant was constructed. The licensee did not initially believe the condition was reportable under 10 CFR 50.72(b)(1)(ii)(B) and 10 CFR 50.73(a)(2)(ii)(B), outside the design basis. However, the staff determined that the condition was reportable under this criterion because the onsite power system lacked suitable redundancy (seismically qualified) as described in GDC 2, GDC 17 and the SAR.[16]

---

[16]   The single failure criterion is discussed in 10 CFR 50 Appendix A, Criterion 17 - Electric Power Systems and the seismic design bases are discussed in 10 CFR 50, Appendix A, Criterion 2 - Design Bases for Protection Against Natural Phenomena, as well as in the FSAR.

3.2.5 External Threat to Plant Safety

§50.72(b)(1)(iii)	§50.73(a)(2)(iii)
Licensee shall report: "Any natural phenomenon or other external condition that poses an actual threat to the safety of the nuclear power plant or significantly hampers site personnel in the performance of duties necessary for the safe operation of the plant."	Licensee shall report: "Any natural phenomenon or other external condition that posed an actual threat to the safety of the nuclear power plant or significantly hampered site personnel in the performance of duties necessary for the safe operation of the nuclear power plant."

If not reported as an emergency under §50.72(a), licensees are required to report any natural phenomenon or other external condition that poses an actual threat to the safety of the nuclear power plant or significantly hampers site personnel in the performance of duties necessary for the safe operation of the plant to the NRC via the ENS as soon as practical and in all cases within 1 hour. Subsequent evaluation may indicate that the phenomenon did not pose an actual threat or significantly hamper site personnel. If so, an LER is not required and the ENS notification may be retracted. Otherwise, licensees are required to submit an LER within 30 days.

Discussion

These criteria apply only to acts of nature (e.g., tornadoes, earthquakes, fires, lightning, hurricanes, floods) and external hazards (i.e., industrial or transportation accidents). References to acts of sabotage are covered by 10 CFR 73.71. Actual threats or significant hampering from internal hazards are covered by separate criteria in §50.72(b)(1)(vi) and §50.73(a)(2)(x), as discussed in Section 3.2.8 of this report.

For ENS reporting, the phrase "actual threat to safety of the nuclear power plant" is one reporting trigger. This covers those events involving an actual threat to the plant from an external condition or natural phenomenon where the threat or damage challenges the ability of the plant to continue to operate in a safe manner (including the orderly shutdown and maintenance of shutdown conditions).

The licensee should decide if a phenomenon or condition actually threatens the plant. For example, a minor brush fire in a remote area of the site that is quickly controlled by fire fighting personnel and, as a result, did not present a threat to the plant should not be reported. However, a major forest fire, large-scale flood, or major earthquake that presents a clear threat to the plant should be reported. As another example, an industrial or transportation accident which occurs near the site, creating a plant safety concern, should be reported.

The licensee must use engineering judgment to determine if there was an actual threat. For example, with regard to tornadoes the decision would be based on such factors as the size of the tornado, and its location and path. There are no prescribed limits. In general, situations involving only monitoring by the plant's staff are not reportable, but if preventive actions are taken or if there are serious concerns, then the situation should be carefully reviewed for reportability.

Responsive actions, by themselves, do not necessarily indicate actual threats. Those which are purely precautionary, such as placement of sandbags, even though flood levels are not expected to be high enough to require sandbags, do not trigger reporting.

Some natural phenomena such as floods may be accurately predicted. If there is a credible prediction of a flood that would challenge the ability of the plant to continue to operate safely, the threat is reportable as an actual threat via ENS as soon as practical and in all cases within 1 hour.

In most cases, events such as earthquakes, approaching hurricanes or tornado warnings result in ENS notification because there is a declaration of an emergency class, which is reportable under §50.72(a)(1)(i) as discussed in Section 3.1.1 of this report, rather than because the event is considered an actual threat. Usually, with the passage of time, it is apparent that an actual threat did not occur and, thus, no LER is submitted (see Example 1). In some cases, with the passage of time, it is judged that an actual threat did occur and, thus, an LER is submitted (see Example 2).

Section 3.2.8 of this report discusses the meaning of the phrase "significantly hampers site personnel in the performance of duties necessary for the safe operation of the plant," in the context of internal threats. A natural phenomenon or external condition, may also significantly hamper personnel. If so, it is reportable under this criterion.

If a snowstorm, hurricane or similar event significantly hampers personnel in the conduct of activities necessary for the safe operation of the plant, the event is reportable via the ENS as soon as practical and in all cases within 1 hour. In the case of snow, the licensee must use judgment based on the amount of snow, the extent to which personnel were hampered, the extent to which additional assistance could have been available in an emergency, the length of time the condition existed, etc. For example, if snow prevented shift relief for several hours, the situation would be reportable if the delay were such that site personnel were significantly hampered in the performance of duties necessary for safe operation. For example, shift personnel might exceed normal shift overtime limits, become excessively fatigued, or find it necessary to operate with fewer than the required number of watchstanders in order to allow some to rest.

<u>Examples</u>

(1)   Earthquake

Seismic alarms were received in the Unit 1 control room of a Southern California plant. Seismic monitors were not tripped in Units 2 or 3. The earthquake was readily felt on site. Seismic instrumentation measured less than 0.02g lateral acceleration.

The licensee classified this as an Unusual Event in accordance with the emergency plan and notified the NRC via ENS per §50.72(a)(1)(i) within 30 minutes of the earthquake. The licensee terminated the event after walkdowns of the plant were satisfactorily completed and made an ENS update call. No LER was submitted because the event was not considered to be an actual threat.

(2)   Hurricane

A licensee in southern Florida declared an Unusual Event after a hurricane warning was issued by the National Hurricane Center. The hurricane was predicted to reach the site in approximately 24 hours. As part of the licensee's severe weather preparations both operating units were taken to hot shutdown before the hurricane's predicted arrival. Offsite power to both units was lost. As the hurricane approached, wind velocity on site was measured in excess of 140 mph. All personnel were withdrawn to protected safety-related structures. Extensive damage occurred on site. The Unusual Event was upgraded to an Alert when the pressurized fire header was lost because of storm-related damage to the fire protection system water supply piping and electric pump. All safety-related equipment functioned as designed before, during, and after the storm with the exception of two minor emergency diesel generator anomalies. The licensee downgraded the Alert to an Unusual Event once offsite power was restored and a damage assessment completed.

An ENS notification was required because the licensee declared an emergency class. The licensee submitted an LER within 30 days of the hurricane, based on the occurrence of a natural phenomenon that posed an actual threat and several other reporting criteria as well.

(3)   Fire

With the unit at 100-percent power, the control room was notified that a forest fire was burning west of the plant close to the 230-kV distribution lines. Approximately 15 minutes later, voltage fluctuations were observed and then a full reactor scram occurred. The licensee determined that the offsite distribution breakers had tripped on fault, apparently from heavy smoke and heat in the vicinity of the offsite 230-kV line insulators. The other source of offsite power, i.e., the 34.5-kV lines supplying the startup transformers, was also lost. Both station emergency diesel generators received a fast start signal and load sequenced as designed. Five minutes later, offsite power was available through the startup transformer to the non-safety-related 4160-v buses, but the licensee decided to maintain the vital buses on their emergency power source until the reliability of offsite power could be assured. The fire continued to burn and, although no plant structures or equipment were directly affected, the fire did approach within 70 feet of the fire pump house.

The licensee entered the emergency plan, declaring an Unusual Event based on high drywell temperature and an Alert based on the potential of the forest fire to further affect the plant. The licensee submitted an LER within 30 days of the fire, based on the occurrence of natural phenomenon that posed an actual threat and several other reporting criteria as well.

3.2.6 ECCS Discharge Into the Reactor Coolant System

§50.72(b)(1)(iv)	10 CFR 50.73
Licensees shall report: "Any event that results or should have resulted in Emergency Core Cooling System (ECCS) discharge into the reactor coolant system as a result of a valid signal."	[ECCS discharge is a subset of §50.73(a)(2)(iv), actuation of an engineered safety feature (ESF), as discussed in Section 3.3.2. Therefore, an LER is required.]

If not reported as an emergency under §50.72(a), licensees are required to notify the NRC via the ENS when a discharge of the ECCS into the RCS occurred or should have occurred as a result of a valid signal as soon as practical and in all cases within 1 hour.

Discussion

Experience with ENS notifications has shown that events involving ECCS discharge to the vessel are generally more serious than ESF actuations without discharge to the vessel. On the basis of this experience, the Commission has made this reporting criterion a 1-hour report. Those events that result in either automatic or manual actuation of the ECCS or would have resulted in activation of the ECCS if some component had not failed or an operator action had not been taken are reportable. For example, if a valid ECCS signal was generated by plant conditions and the operator put all ECCS pumps in pull-to-lock position, although no ECCS discharge occurred, the event is reportable.

A "valid signal" refers to the actual plant conditions or parameters satisfying the requirements for ECCS initiation. Valid actuations also include intentional manual actuations, unless the actuation is part of a preplanned sequence during test or operation. Excluded from this reporting requirement would be those instances in which instrument drift, spurious signals, human error, or other invalid signals caused actuation of the ECCS (e.g., jarring a cabinet, an error in the use of jumpers or lifted leads, an error in the actuation of switches or controls, equipment failure or radio frequency interference). However, such events may be reportable under other criteria; in particular, if an ESF is actuated §50.72(b)(2)(ii) requires a report within four hours and §50.73(a)(2)(iv) requires submittal of an LER.

The staff considers deliberate manual ECCS initiations or actuations based on the operator's understanding of actual plant conditions or parameters as valid signals. However, inadvertent manual ECCS initiations or actuations that occur because of human error, such as errors that occur during surveillance tests or maintenance activities, are not considered as valid signals. If the ECCS discharged or should have discharged into the reactor coolant system as a result of an invalid signal, no ENS notification under this reporting criterion is required. (Such a condition may be reportable as an ESF actuation under 10 CFR 50.72(b)(2)(ii).)

Any event reportable under §50.72(b)(1)(iv) also requires a 30-day LER under §50.73(a)(2)(iv) because an ESF was actuated.

Examples

(1)   BWR Scram and ECCS Injection on Valid Signal

A loss of instrument air caused the feedwater pump minimum flow valves to fail open and decrease reactor vessel level. This resulted in an automatic reactor scram/turbine trip and high-pressure core spray and reactor core isolation cooling injection into the reactor vessel for 4 minutes. After reactor vessel level and the condensate and feedwater systems were restored, these pumps were secured.

An ENS notification is required under §50.72(b)(1)(iv) because an ECCS system injected water into the RCS as a result of a valid ECCS signal. Although the RPS actuation also is reportable within 4 hours under §50.72(b)(2)(ii), this more limiting criterion applies. An LER is required under §50.73(a)(2)(iv) because an ESF actuation occurred.

(2)   PWR ECCS Injection following Surveillance Testing

While making preparations for a normal plant cooldown in Mode 5, the licensee performed stroke time testing of the safety injection isolation valves. Following the test these valves were not returned to the closed position. This resulted in approximately 2000 gallons of borated water injecting into the reactor coolant system when the plant was depressurized below the safety injection tank pressure of 260 psia.

This event is reportable as an ECCS injection under §50.72(b)(1)(iv). ECCS initiation was based on RCS pressure being less than safety injection tank pressure. Therefore, ECCS initiation is considered to result from a valid signal. An LER is required under §50.73(a)(2)(iv).

(3)   PWR ECCS Injection Caused by Personnel Error

While surveillance testing containment isolation valves, a test push-button was inadvertently released, which initiated a "B" train containment isolation and ECCS. High-pressure ECCS pumps injected 300 gallons of borated water from the refueling water storage tank into the reactor before the "B" pumps were secured while the reactor remained at 94-percent power.

This event is not reportable under §50.72(b)(1)(iv), even though it was an ECCS injection into the RCS, because it resulted from an invalid signal; however, it is reportable as an ESF actuation under §50.72(b)(2)(ii) and an LER is required under §50.73(a)(2)(iv).

3.2.7 Loss of Emergency Preparedness Capabilities

§50.72(b)(1)(v)	10 CFR 50.73
Licensees shall report: "Any event that results in a major loss of emergency assessment capability, offsite response capability, or communications capability (e.g., significant portion of control room indication, Emergency Notification System, or offsite notification system)."	[No corresponding Part 50.73 requirement.]

If not reported as an emergency under 50.72(a), licensees are required to notify the NRC of a major loss of their emergency assessment, offsite response, or communications capability as soon as practical and in all cases within 1 hour.

Discussion

This reporting requirement pertains to events that would impair a licensee's ability to deal with an accident or emergency. Notifying the NRC of these events may permit the NRC to take some compensating measures and to more completely assess the consequences of such a loss should it occur during an accident or emergency.

Examples of events that this criterion is intended to cover are those in which any of the following is not available:

- Safety parameter display system (SPDS)

- Emergency response facilities (ERFs)

- Emergency communications facilities and equipment including the emergency notification system (ENS)

- Public prompt notification system including sirens

- Plant monitors necessary for accident assessment

These and other situations should be evaluated for reportability as discussed below.

Loss of Emergency Assessment Capability

A major loss of emergency assessment capability would include those events that significantly impair the licensee's safety assessment capability. Some engineering judgment is needed to determine the significance of the loss of particular equipment, e.g., loss of only the SPDS for a short period of time need not be reported, but loss of SPDS and other assessment equipment at the same time may be reportable.

The staff considers the loss of a significant portion of control room indication including annunciators or monitors, or the loss of all plant vent stack radiation monitors, as examples of a major loss of emergency assessment capability which should be evaluated for reportability.

Loss of Offsite Response Capability

A major loss of offsite response capability includes those events that would significantly impair the fulfillment of the licensee's approved emergency plan for other than a short time. Loss of offsite response capability may typically include the loss of plant access, emergency offsite response facilities[17], or public prompt notification system, including sirens and other alerting systems.

If a ~~large storm~~ significant natural hazard (e.g., earthquake, hurricane, tornado, flood, etc.) or other event causes ~~roads to be closed~~ evacuation routes to be impassible or other parts of the response infrastructure to be impaired to the extent that the State and local governments are rendered incapable of fulfilling their responsibilities in the emergency plan for the plant, then the NRC must be notified. This does not apply in the case of routine traffic impediments such as fog, snow and ice which do not render the state and local governments incapable of fulfilling their responsibilities. It is intended to apply to more significant cases such as the conditions around the Turkey Point plant after Hurricane Andrew struck in 1992 or the conditions around the Cooper station during the Midwest floods of 1993.

If the alert systems, e.g., sirens, are owned and/or maintained by others, the licensee should take reasonable measures to remain informed and must notify the NRC if a large number of sirens fail. Although the loss of a single siren for a short time is not a major loss of offsite response capability, the loss of a large number of sirens, other alerting systems (e.g., tone alert radios), or more importantly, the lost capability to alert a large segment of the population for 1 hour would warrant an immediate notification.

Loss of Communications Capability

A major loss of communications capability may include the loss of ENS and/or other offsite communication systems. The other offsite communication systems may include a dedicated telephone communication link to a State or a local government agency and emergency offsite response facilities, in-plant paging and radio systems required for safe plant operation, or commercial telephone lines.

Should either or both of the emergency communications subsystems (ENS and HPN) fail, the NRC Operations Center should be so informed over normal commercial telephone lines. When notifying the NRC Operations Center, licensees should use the backup commercial telephone

---

[17] Performing preventive maintenance on an offsite emergency response facility is not reportable if the facility can be returned to service promptly in the event of an accident.

numbers provided. This satisfies the guidance provided in previous Information Notices 85-44 "Emergency Communication System Monthly Test," dated May 30, 1985 and 86-97 "Emergency Communications System," dated November 28, 1986, to test the backup means of communication when the primary system is unavailable as well as the reporting requirements of §50.72(b)(1)(v). If the Operations Center notifies the licensee that an ENS line is inoperable, there is no need for a subsequent licensee notification. Loss of either ENS or HPN does not generate an event report. The Operations Center contacts the appropriate repair organization.

In a similar manner, if the NRC supplied telephone line or modem used for the emergency response data system is inoperable, the NRC operations center should be informed so that repairs can be ordered. However, this does not generate an event report.

Examples

(1)   Plant Access Roads Closed by Storm

The local sheriff notified the licensee that all roads to and from the plant were closed because of a snow storm. The licensee had two full-shift crews on site to support plant operations and no emergency declaration was made. The licensee notified State and local authorities of the situation and made an ENS notification. The licensee deactivated its station isolation procedures after the storm passed and the roads were passable.

An ENS notification was made because the licensee determined that the road closing constituted a major loss of emergency offsite response capability. No LER is required.

(2)   Loss of Public Prompt Notification System

ENS notifications of the loss of the emergency sirens or tone alert radios vary according to the licensee's locale and interpretations of "major loss" and have included:

*   12 of 40 county alert sirens disabled because of loss of power as a result of severe weather.

*   28 of 54 alert sirens were reported out of service as a result of a local ice storm.

*   All offsite emergency sirens were:

    - found inoperable during a monthly test:
    - taken out of service for repair.
    - inoperable because control panel power was lost.
    - inoperable because the county radio transmitter failed.

An ENS notification is required because of the major loss of offsite response capability, i.e., the public prompt notification system. However, licensees may use engineering judgment in determining reportability (i.e., a "major loss") based upon such factors as the percent of the population not covered by emergency sirens and the existence of procedures or practices to compensate for the lost emergency sirens. An LER is not required because there is no corresponding 10 CFR 50.73 requirements.

(3)   Loss of ENS and Commercial Telephone System

The licensee determined that ENS and commercial telecommunications capability was lost to the control room when a fiber optic cable was severed during maintenance. A communications link was established and maintained between the site and the load dispatcher via microwave transmission. Both the ENS and commercial communications capability were restored approximately 90 minutes later.

An ENS notification is required because of the major loss of communications capability. Although the microwave link to the site was established and maintained during the telephone outage, this in itself does not fully compensate for the loss of communication that would be required in the event of an emergency at the plant. No LER is required because there is no corresponding 10 CFR 50.73 requirements.

(4)   Loss of Direct Communication Line to Police

The licensee contacted the State Police via commercial telephone lines and reported to the NRC Operations Center that the direct telephone line to the State Police was inoperable for over 1 hour. The licensee notified the NRC Operations Center in a followup ENS call that the line was restored to operability.

An ENS notification would be required if the loss of the direct telephone line(s) to various police, local, or State emergency or regulatory agencies is not compensated for by other readily available offsite communications systems. In this example, no ENS notification is required since commercial telephone lines to the State Police were available. No LER is required because there is no corresponding 10 CFR 50.73 requirements.

3.2.8 Internal Threat to Plant Safety

§50.72(b)(1)(vi)	§50.73(a)(2)(x)
Licensees shall report: "Any event that poses an actual threat to the safety of the nuclear power plant or significantly hampers site personnel in the performance of duties necessary for the safe operation of the nuclear power plant including fires, toxic gas releases, or radioactive releases."	Licensees shall report: "Any event that posed an actual threat to the safety of the nuclear power plant or significantly hampered site personnel in the performance of duties necessary for the safe operation of the nuclear power plant including fires, toxic gas releases, or radioactive releases."

If not reported as an emergency under §50.72(a), licensees are required to report such an event or condition to the NRC via the ENS as soon as practical and in all cases within 1 hour. Licensees are required to submit an LER within 30 days.

Discussion

These criteria pertain to internal threats. The criteria for external threats, §50.72(b)(1)(iii) and §50.73(a)(2)(iii), are described in Section 3.2.5.

This provision requires reporting events, particularly those caused by acts of personnel, which endanger the safety of the plant or interfere with personnel in the performance of duties necessary for safe plant operations.

The licensee must exercise some judgment in reporting under this rule. For example, a small fire on site that did not endanger any plant equipment and did not and could not reasonably be expected to endanger the plant is not reportable.

As indicated in the Statement of Considerations the phrase "significantly hampers site personnel" applies narrowly, i.e. only to those events which significantly hamper the ability of site personnel to perform safety-related activities affecting plant safety.[18]

In addition, the staff considers the following standards appropriate in this regard:

- The significant hampering criterion is pertinent to "the performance of duties necessary for safe operation of the nuclear power plant." One way to evaluate this is to ask if one could seal the room in question (or disable the function in question) for a substantial period of time and still operate the plant safely. For example, if a switchgear room is unavailable for a time, but it is normally not necessary to enter the room for safe operation, and no need to enter the room arises while it is unavailable, the event is not reportable under this criterion.

---

[18]   48 FR 33856, July 26, 1983.

- Significant hampering includes hindering or interfering (such as with protective clothing or radiation work permits) provided that the interference or delay is sufficient to significantly threaten the safe operation of the plant.

- Actions such as room evacuations that are precautionary would not constitute significant hampering if the necessary actions can still be performed in a timely manner.

Plant mode may be considered in determining if there is an actual internal threat to a plant. However, licensees should not incorrectly assume that everything that happens while a plant is shut down is unimportant and not reportable.

In-plant releases must be reported if they require evacuation of rooms or buildings ~~containing systems important to safety~~ and, as a result, the ability of the operators to perform necessary duties is significantly hampered.

Events such as minor spills, small gaseous waste releases, or the disturbance of contaminated particulate matter (e.g., dust) that require temporary evacuation of an individual room until the airborne concentrations decrease or until respiratory protection devices are used, are not reportable unless the ability of site personnel to perform necessary safety functions is significantly hampered.

No LER is required for precautionary evacuations of rooms and buildings that subsequent evaluation determines were not required. Even if an evacuation affects a major part of the facility, the test for reportability is whether an actual threat to plant safety occurred or whether site personnel were significantly hampered in carrying out their safety responsibilities.

In most cases, fires result in ENS notification because there is a declaration of an emergency class, which is reportable under §50.72(a)(1)(ii) as discussed in Section 3.1.1 of this report.[19] If there is an actual threat or significant hampering, an LER is also required. With regard to control room fires, the staff generally considers a control room fire to constitute an actual threat and significant hampering.[20]

---

[19] As indicated in NUREG-0654, Rev. 1, Information Notice 88-64 and Regulatory Guide 1.101, Rev. 3 (which endorses NUMARC/NESP-007, Rev. 2), a fire that lasts longer than 10 or 15 minutes or which affects plant equipment important for safe operation would result in declaration of an emergency class.

[20] It is theoretically possible to have a control room fire which is discovered and extinguished quickly and, even in this location, does not significantly hamper the operators and does not threaten plant safety. Examples could include small paper fires in ash trays or trash cans, or cigarette burns of furniture or upholstery.

Examples

(1)   Fires

*   Question:

If we have a fire in the refueling bridge and we are not moving fuel, would the fire be reportable?

Answer:

No. If the plant is not moving fuel and the fire does not otherwise threaten other safety equipment and does not hamper site personnel, the fire is not reportable. If the plant is moving fuel, the fire is reportable.

*   Question:

If we have a fire in the reactor building that forces contractor personnel who are doing a safety related modification to leave, but the fire did not hamper operations personnel or equipment, would that fire be reportable?

Answer:

No. The fire would not be reportable if the fire was not severe enough that it posed an actual threat to the plant and the delay in completing the modification did not significantly threaten the safe operation of the plant.

3.3 Four-hour ENS Notifications and LERs

This section addresses §50.72(b)(2), "Non-Emergency Events--Four-Hour Reports," and 10 CFR 50.73 written reports associated with these 50.72 notifications. If not reported as a declaration of emergency class under §50.72(a) or as a non-emergency 1-hour report under §50.72(b)(1), licensees are to notify the NRC as soon as practical and in all cases within 4 hours of the occurrence of any of the events required by §50.72(b)(2) and to submit an LER within 30 days for any event or condition required by 10 CFR 50.73.

In addition to events reportable under both 10 CFR 50.72 and 50.73, several requirements for 50.72 notifications only or LERs only are included in this section because of the sequential numbering scheme used. For example, common-mode failures of channels, trains, or systems, as discussed in Section 3.3.4, require LERs, but no ENS notifications are explicitly required unless reportable under other criteria. Transport of a contaminated person to an offsite medical facility, as discussed in Section 3.3.6, requires ENS notification but no LER.

3.3.1 Shutdown Plant Found in Degraded or Unanalyzed Condition

§50.72(b)(2)(i)	10 CFR 50.73
Licensees shall report: "Any event <u>found while the reactor is shut down</u>, that, <u>had it been found while the reactor was in operation, would have</u> resulted in the nuclear power plant, including its principal safety barriers, being seriously degraded or being in an unanalyzed condition that significantly compromises plant safety."	[Events found while the reactor is shutdown that involve degradation of the principal safety barriers or unanalyzed conditions that significantly compromise plant safety are addressed by §50.73(a)(2)(ii). Therefore, an LER is required. See Section 3.2.4.]

If not reported under §50.72(a) or (b)(1), licensees are required to report any such condition to the NRC via the ENS as soon as practical, and in all cases within 4 hours of discovery of the condition. Licensees are required to submit an LER within 30 days.

Discussion

Guidelines for identifying events that would result in the nuclear power plant being seriously degraded or being in an unanalyzed condition that significantly compromises plant safety are discussed in Section 3.2.4 of this report.

Examples

(1)   Significant Degradation of Reactor Fuel Rod Cladding Identified During Testing of Fuel Assemblies

Radio-chemistry data for a particular PWR indicated that a number of fuel rods had failed during the first few months of operation. Projections ranged from 6 to 12 failed rods. The end of cycle reactor coolant system iodine-131 activity averaged 0.025 micro curies per milliliter. following the end of cycle shutdown, iodine-131 spiked to 11.45 micro curies per milliliter. The cause was due to a significant number of failed fuel rods. Inspections revealed that 136 of the total 157 fuel assemblies contained failed fuel (approximately 300 fuel rods had through-wall penetrations), far exceeding the anticipated number of failures. The defects were generally pinhole sized. The fuel cladding failures were caused by long-term fretting from debris that became lodged between the lower fuel assembly nozzle and the first spacer grid, resulting in penetration of the stainless-steel fuel cladding. The source of the debris was apparently a machining byproduct from the thermal shield support system repairs during the previous refueling outage.

An ENS notification is required because a principal safety barrier (the fuel cladding) was found seriously degraded. An LER is required.

(2)   **Corrosion of a Control Rod Drive Mechanism Flange Resulted in a Reactor Coolant System Pressure Boundary Degradation**

While the plant was in hot shutdown, a total of six control rod drive mechanism (CRDM) reactor vessel nozzle flanges were identified as leaking. Subsequently one of the flanges was found eroded and pitted. While removing the nut ring from beneath the flange, it was discovered that approximately 50 percent of one of the nut ring halves had corroded away and that two of the four bolt holes in the corroded nut ring half were degraded to the point where there was no bolt/thread engagement.

An inspection of the flanges and spiral wound gaskets, which were removed from between the flanges, revealed that the cause of the leaks was the gradual deterioration of the gaskets from age. A replacement CRDM was installed and the gaskets on all six CRDMs were replaced with new design graphite-type gaskets.

An ENS notification is required because the condition caused a significant degradation of the RCS pressure boundary. An LER is required.

(3)   **Significant Degradation of Reactor Fuel Rod Cladding Identified During Fuel Sipping Operations**

With the plant in cold shutdown, fuel sipping operations identified a significant portion of cycle 2 fuel, type "LYP," had failed, i.e., four confirmed and twelve potential fuel leakers. The potential fuel leakers had only been sipped once prior to making the ENS notification. The licensee contacted the fuel vendor for assistance on-site in evaluating this problem.

As in example (1), an ENS notification was made because a principal safety barrier (the fuel cladding) was found seriously degraded. However, additional sipping operations and a subsequent evaluation by the licensee's reactor engineering department with vendor assistance concluded that no additional fuel failures had occurred, i.e., the abnormal readings associated with the potential fuel leakers was attributed to fission products trapped in the crud layer. Based on the results of the evaluation the licensee concluded that the fuel cladding was not seriously degraded and that the event was not reportable. Consequently, after discussion with the Regional Office, the licensee retracted this event.

3.3.2 Actuation of an Engineered Safety Feature or the RPS

§50.72(b)(2)(ii)	§50.73(a)(2)(iv)

Licensees shall report "any event or condition that results in a manual or automatic actuation of any Engineered Safety Feature (ESF), including the Reactor Protection System (RPS) except when:
  (A) The actuation results from and is part of the preplanned sequence during testing or reactor operation;
  (B) The actuation is invalid and:
    (1) Occurs while the system is properly removed from service;
    (2) Occurs after the safety function has been already completed; or
    (3) Involves only the following specific ESFs or their equivalent systems;
      (i) Reactor water clean-up system;
      (ii) Control room emergency ventilation system;
      (iii) Reactor building ventilation system;
      (iv) Fuel building ventilation system; or
      (v) Auxiliary building ventilation system."
	Licensees shall report "any event or condition that resulted in a manual or automatic actuation of any Engineered Safety Feature (ESF), including the Reactor Protection System (RPS), except when:
	  (A) The actuation resulted from and was part of a pre-planned sequence during testing or reactor operation;
	  (B) The actuation was invalid and:
	    (1) Occurred while the system was properly removed from service;
	    (2) Occurred after the safety function had been already completed; or
	    (3) Involved only the following specific ESFs or their equivalent systems;
	      (i) Reactor water clean-up system;
	      (ii) Control room emergency ventilation system;
	      (iii) Reactor building ventilation system;
	      (iv) Fuel building ventilation system; or
	      (v) Auxiliary building ventilation system."

If not reported under §50.72(a) or (b)(1), licensees are required to report any engineered safety feature actuation, including the reactor protection system, to the NRC via the ENS as soon as practical and in all cases within 4 hours of the event. Licensees are required to submit an LER within 30 days.

Discussion

The Statements of Considerations indicate that this paragraph requires events to be reported whenever an ESF actuates either manually or automatically, regardless of plant status. It is based on the premise that the ESFs are provided to mitigate the consequences of a significant event and, therefore: (1) they should work properly when called upon, and (2) they should not be challenged frequently or unnecessarily. The Commission is interested both in events where an ESF was needed to mitigate the consequences (whether or not the equipment performed properly) and events where an ESF actuated unnecessarily. In discussing the reporting of actuations which are part of preplanned procedures, the Statements of Considerations also state that actuations that need not be reported are those initiated for reasons other than to mitigate the consequences of an event (e.g., at the discretion of the licensee as part of a preplanned procedure).[21]

This indicates an intent to require reporting actuations of features that mitigate the consequences of significant events. Usually, the staff would not consider this to include single component actuations because single components of complex systems, by themselves, usually do not mitigate the consequences of significant events. However, in some cases a component would be sufficient to mitigate the event (i.e., perform the ESF function) and its actuation would, therefore, be reportable. This position is consistent with the statement that the reporting requirement is based on the premise that ESFs are provided to mitigate the consequences of a significant event.

Single trains do mitigate the consequences, and, thus, train level actuations are reportable.

In this regard, the staff considers actuation of a diesel-generator to be actuation of a train—not actuation of a single component — because a diesel generator mitigates the event (performs the ESF function for plants at which diesel generators are classified as ESF systems). (See Example 3 below.)

The staff also considers intentional manual actions, in which one or more ESF components are actuated in response to actual plant conditions resulting from equipment failure or human error, to be reportable because such actions would usually mitigate the consequences of a significant event. This position is consistent with the statement that the Commission is interested in events where an ESF was needed to mitigate the consequences of the event. For example, starting a safety injection pump in response to a rapidly decreasing pressurizer level or starting HPCI in response to a loss of feedwater would be reportable. However, shifting alignment of makeup pumps or closing a containment isolation valve for normal operational purposes would not be reportable.

The Statement of Considerations also indicates that "actuation" of multichannel ESF actuation systems is defined as actuation of enough channels to complete the minimum actuation logic. Therefore, single channel actuations, whether caused by failures or otherwise, are not reportable if they do not complete the minimum actuation logic.[22] Note, however, that if only a single logic channel actuates when, in fact, the ESF system should have actuated in response to plant parameters, this would be reportable as an ESF failure. The event would be reportable under these criteria (ESF actuation) as well as under 10 CFR 50.72(b)(2)(iii) and 10 CFR 50.73(a)(2)(v) (event or condition alone). This position is consistent with the statement that the Commission is interested in events where an ESF was needed to mitigate the consequences, whether or not the equipment performed properly.[23]

---

[21] 48 FR 33854, July 28, 1983, 48 FR 39043 and 48 FR 39044, August 29, 1983.

[22] 48 FR 33854, July 28, 1983, 48 FR 39043 and 48 FR 39044, August 29, 1983.

[23] Also see 48 FR 39043, August 29, 1983, which states that this paragraph is intended to capture events during which an ESF actuates or fails to actuate.

With regard to preplanned actuations, the Statements of Consideration indicate that operation of an ESF as part of a planned test or operational evolution need not be reported. Preplanned actuations are those which are expected to actually occur due to preplanned activities covered by procedures. Such actuations are those for which a procedural step or other appropriate documentation indicates the specific ESF actuation that is actually expected to occur. Control room personnel are aware of the specific signal generation before its occurrence or indication in the control room. However, if during the test or evolution, the ESF actuates in a way that is not part of the planned evolution, that actuation should be reported. For example, if the normal reactor shutdown procedure requires that the control rods be inserted by a manual reactor scram, the reactor scram need not be reported. However, if unanticipated conditions develop during the shutdown that cause an automatic reactor scram, such a reactor scram should be reported. The fact that the safety analysis assumes that an ESF will actuate automatically during an event does not eliminate the need to report that actuation. Actuations that need not be reported are those initiated for reasons other than to mitigate the consequences of an event (e.g., at the discretion of the licensee as part of a planned evolution).[24]

Note that if an operator were to manually scram the reactor in anticipation of receiving an automatic reactor scram, this would be reportable just as the automatic scram would be reportable.

On September 10, 1992, the Commission published final amendments to 10 CFR 50.72 and 50.73 that apply to reporting of ESF actuations. Three categories of invalid ESF actuations are not reportable. These three categories are invalid ESF actuations of (1) systems which had been properly removed from service, or (2) systems for which the safety function which the ESF is intended to accomplish had already been accomplished, and (3) several specific systems listed below.

Valid ESF actuations are those actuations that result from "valid signals" or from intentional manual initiation, unless it is part of a preplanned test. Valid signals are those signals that are initiated in response to actual plant conditions or parameters satisfying the requirements for ESF initiation. Note this definition of "valid" requires that the initiation signal must be an ESF signal. This distinction eliminates actuations which are the result of non-ESF signals from the class of valid actuations. Invalid actuations are, by definition, those that do not meet the criteria for being valid. Thus, invalid actuations include actuations that are not the result of valid signals and are not intentional manual actuations.

Invalid ESF actuations that occur when the system is already properly removed from service are not reportable if all requirements of plant procedures for removing equipment from service have been met. This includes required clearance documentation, equipment and control board tagging, and properly positioned valves and power supply breakers. In addition, invalid ESF actuations that occur after the safety function has already been completed are not reportable An example would be RPS actuation after the control rods have already been inserted into the core.

---

[24]   48 FR 33854, July 28, 1983, 48 FR 39043 and 48 FR 39044, August 29, 1983.

Finally, invalid actuations for several specific systems or their equivalent are not reportable. These systems are the reactor water clean up system in boiling water reactors (BWRs), the control room emergency ventilation system, the reactor building ventilation system (RBVS), the fuel building ventilation system, and the auxiliary building ventilation system. Thus, reporting of invalid actuations for these specific systems due to signals that originated from non-ESF circuitry are not required.

Invalid actuations of other ESF systems continue to be reportable. For BWRs, the actuation of the standby gas treatment system following an invalid actuation of the RBVS is also not reportable.

If an invalid ESF actuation reveals a defect in the ESF system so the system failed or would fail to perform its intended function, the event continues to be reportable under other requirements of 10 CFR 50.72 and 50.73. When invalid ESF actuations excluded by the conditions described above occur as part of a reportable event, they should be described as part of the reportable event, in order to provide a complete, accurate and thorough description of the event.

The reporting criterion "is based on the premise that ESFs are provided to mitigate the consequences of a significant event ..."[25] Systems typically reported under this criterion include the systems listed in Table 2. These are systems required to mitigate significant events and include ECCS, RPS, containment systems and certain auxiliary and support systems required to perform ESF functions. These are systems that are described in the FSAR and are required to satisfy ESF functional requirements. The NRC staff considers these systems to be a reasonable interpretation of what constitutes systems "provided to mitigate the consequences of a significant event."

<u>Examples</u>

(1)   RPS Actuation

- The licensee was placing the residual heat removal (RHR) system in its shutdown cooling mode while the plant was in hot shutdown. The BWR vessel level decreased for unknown reasons, causing a RPS scram and Group III primary containment isolation signals, as designed. All control rods had been previously inserted and all Group III isolation valves had been manually isolated. The licensee isolated RHR to stop the decrease in reactor vessel level.

  This event is reportable within 4 hours under this criterion because, although the systems' safety functions had already been completed, the RPS scram and primary containment isolation signals were valid and the actuations were not part of the planned procedure. The automatic signals were valid because they were generated from the sensor by measurement of an actual physical system parameter that was at its set point. An LER is required.

---

[25]   48 FR 33854, July 26, 1983.

- With the BWR defueled, an invalid signal actuated the RPS. There was no component operation because the control rod drive system had been properly removed from service. This event is not reportable because (1) the RPS signal was invalid, _and_ (2) the system had been properly removed from service.

- An immediate notification (§50.72) was received from a BWR licensee. In the reported event, both recirculation pumps tripped as a result of a breaker problem. This placed the plant in a condition in which BWRs are generally scrammed to avoid potential power/flow oscillations. At this plant, for this condition, a written off-normal procedure required the plant operations staff to scram the reactor. The plant staff performed a reactor scram which was uncomplicated. This event is reportable as a manual RPS actuation. Even though the reactor scram was in response to an existing written procedure, this event does not involve a preplanned sequence because the loss of recirculation pumps and the resultant off-normal procedure entry were event driven, not preplanned. An LER is required. In this case, the licensee initially retracted the ENS notification believing that the event was not reportable. After staff review and further discussion, it was agreed that the event is reportable for the reasons discussed above.

(2)   BWR Control Rod Block Monitor Actuation

A rod block that was part of the planned startup procedure occurred from the rod block monitor, which, at this plant, is classified as a portion of the RPS or as an ESF.

This event is not reportable because it occurred as a part of a preplanned startup procedure that specified certain rod blocks were expected to occur.

(3)   Emergency Diesel Generator (EDG) Starts

- The licensee provided an LER describing an event in which the EDG automatically started when a technician inadvertently caused a short circuit that de-energized an essential bus during a calibration. An ENS notification and LER are required because the EDG auto-start (ESF actuation at this plant) was not identified at the step in the calibration procedure being used.

- The licensee provided an LER describing an event in which, after an automatic EDG start, and for unknown reasons, the emergency bus feeder breaker from the EDG did not close when power was lost on the bus. An ENS notification and LER are required because the actuation logic for the EDG start (ESF actuation at this plant) was completed, even though the diesel generator did not power the safety buses.

(4)   Preplanned Manual Scram

During a normal reactor shutdown, the reactor shutdown procedure required that reactor power be reduced to a low power at which point the control rods were to be inserted by a manual reactor scram. The rods were manually scrammed.

This event is not reportable because the manual scram results from and is, by procedure, part of a preplanned sequence of reactor operation. However, if conditions develop during the process of shutting down that require an unplanned reactor scram, the RPS actuation (whether manually or automatically produced) is reportable via ENS notification and LER.

(5)   Actuation of Wrong Component During Testing

During surveillance testing of the main steam isolation valves (MSIVs), an operator incorrectly closed MSIV "D" when the procedure specified closing MSIV "C."

This event is not reportable because the event is an inadvertent actuation of a single component of an ESF system rather than a train level actuation (and the purpose of the actuation was not to mitigate the consequences of an event).

(6)   Control Room Ventilation System (CRVS) Isolation

While the CRVS was in service with no testing or maintenance in progress, a voltage transient caused spiking of a radiation monitor resulting in isolation of the CRVS, as designed.

This event is not reportable under this criterion because the event is due to an invalid signal <u>and</u> involves one of the four excepted systems (CRVS).

(7)   Reactor Water Cleanup (RWCU) Isolations

- The RWCU isolation valves closed in response to high water temperature, as designed. This is a common operational occurrence not indicative of a significant event; the initiation signal for this isolation is a non-ESF signal. As discussed above, this is an invalid actuation because it originates from a non-ESF signal and the event is not reportable because it is an invalid actuation of one of the four excepted systems.

- An RWCU primary containment isolation (ESF actuation) occurred on pressurization between the RWCU suction containment isolation valves during the restoration of the RWCU system after a maintenance outage. An ENS notification and LER are required because a valid ESF signal initiated the RWCU isolation and the actuation was not part of a planned procedure.

(8)   Manual Actuation of ESF Component in Response to Actual Plant Condition

At a PWR, maintenance personnel inadvertently pulled an instrument line out of a compression fitting connection at a pressure transmitter. The resultant reactor coolant system (RCS) leak was estimated at between 70 and 80 gpm. Charging flow increased due to automatic control system action. The operations staff recognized the symptoms of an RCS leak and entered the appropriate off-normal procedure. The procedure directed the operations staff to start a second charging pump and flow was manually increased to raise pressurizer level. Based on the response of the pressurizer level, the operations staff determined that a reactor scram and safety injection were not necessary. Maintenance personnel still at the transmitter closed the instrument block and root valves terminating the event.

The staff considers the manual start of the charging pump (which also serves as an ECCS pump, but with a different valve lineup) in response to dropping pressurizer level to be an intentional manual actuation of an ESF in response to equipment failure or human error and reportable because it constitutes deliberate manual actuation of a single component of an ESF, in response to plant conditions, to mitigate the consequences of an event. As indicated in the Statements of Considerations for the rules "Actuations that need not be reported are those that are initiated for reasons other than to mitigate the consequences of an event (e.g., at the discretion of the licensee as part of a planned procedure or evolution)."[26]

(9)   ESF Actuation During Maintenance Activity

At a BWR, a maintenance activity was under way involving placement of a jumper to avoid ESF actuations. The maintenance staff recognized that there was a high potential for a loss of contact with the jumper and consequent ESF actuation. This potential was explicitly stated in the maintenance work request and on a risk evaluation sheet. The operating staff was briefed on the potential ESF actuations prior to start of work. During the event, a loss of continuity did occur and the ESF actuations involving isolation, standby gas treatment start, closing of some valves in the primary containment isolation system (recirculation pump seal mini-purge valve, nitrogen supply to drywell valve, and containment atmospheric monitoring valve) occurred.

The staff has concluded that the event would not be reportable if the event were described in appropriate documentation as definitely expected to occur. However, since the event was not listed as definitely expected to occur and was not an intended result of the planned procedure, the event is reportable.

---

[26]   48 FR 39043, August 29, 1983, and 48 FR 33854, July 26, 1983.

Table 2. Example Systems

Emergency Core Cooling Systems (ECCS) for Pressurized Water Reactors (PWRs):
    • reactor coolant system accumulators
    • boron injection system
    • high-, intermediate-, and low-head injection systems, including systems for charging using centrifugal charging pumps, safety injection systems, and residual (decay) heat removal systems
 	
ECCS for Boiling Water Reactors (BWRs):
    • high- and low-pressure core spray systems
    • high-pressure coolant injection system, feedwater coolant injection system, residual heat removal system (low pressure injection portion)
    • isolation condenser system, reactor core isolation cooling system
    • automatic depressurization system

Containment Systems
    • containment and reactor vessel isolation systems
    • containment heat removal and depressurization systems, including the containment spray and additive system and the fan cooler system
    • containment air purification and cleanup systems
    • containment combustible gas control systems, including hydrogen recombiners, igniters, and containment atmospheric dilution systems
    • BWR standby gas treatment systems

Electrical Systems
    • emergency ac electrical power systems, including emergency diesel generators (EDGs) and their associated support systems and BWR dedicated Division 3 EDGs and their associated support systems
    • actuation and control systems

Heating, Ventilating and Air Conditioning (HVAC) Systems for Control Room and Fuel Handling Areas
Anticipated Transient Without Scram (ATWS) Mitigating Systems
PWR Auxiliary Feedwater Systems

3.3.3 Event or Condition That Alone Could Prevent Fulfillment of a Safety Function

§50.72(b)(2)(iii)	§50.73(a)(2)(v)
Licensees shall report: "Any event or condition that alone could have prevented the fulfillment of the safety function of structures or systems that are needed to: (A) Shut down the reactor and maintain it in a safe shutdown condition; (B) Remove residual heat; (C) Control the release of radioactive material; or (D) Mitigate the consequences of an accident."	Licensees shall report: "Any event or condition that alone could have prevented the fulfillment of the safety function of structures or systems that are needed to: (A) Shut down the reactor and maintain it in a safe shutdown condition; (B) Remove residual heat; (C) Control the release of radioactive material; or (D) Mitigate the consequences of an accident."
10 CFR 50.72	**§50.73(a)(2)(vi)**
[The Statements of Consideration for 10 CFR 50.72 contain wording similar to those of §50.73(a)(2)(vi).]	"Events covered in paragraph (a)(2)(v) of this section may include one or more personnel errors, equipment failures, and/or discovery of design, analysis, fabrication, construction, and/or procedural inadequacies. However, individual component failures need not be reported pursuant to this paragraph if redundant equipment in the same system was operable and available to perform the required safety function".

If not reported under §50.72(a) or (b)(1), licensees shall notify the NRC via the ENS as soon as practical and in all cases within 4 hours of discovery of the event or condition and submit an LER within 30 days.

Discussion

The level of judgment for reporting an event or condition under this criterion is a reasonable expectation of preventing fulfillment of a safety function. In the discussions which follow, many of which are taken from the Statement of Considerations or from previous NUREG guidance, several different expressions such as "would have," "could have," "alone could have," and "reasonable doubt" are used to characterize this standard. In the staff's view, all of these should be judged on the basis of a reasonable expectation of preventing fulfillment of the safety function.

As indicated in the Statement of Considerations, the intent of these criteria is to capture those events where there would have been a failure of a safety system to properly complete a safety function, regardless of when the failures were discovered or whether the system was needed at the time.[27]

These criteria cover an event or condition where ~~redundant~~ structures, components, or trains of a safety system could have failed to perform their intended function because of: one or more personnel errors, including procedure violations; equipment failures; inadequate maintenance; or design, analysis, fabrication, equipment qualification, construction, or procedural deficiencies. The event must be reported ~~regardless of the situation or condition that caused the structure or systems to be unavailable, and~~ regardless of whether or not an alternate safety system could have been used to perform the safety function (e.g., high pressure core cooling failed, but feed-and-bleed or low pressure core cooling were available to provide the safety function of core cooling).

The definition of the systems included in the scope of these criteria is provided in the rules themselves; it is not determined by the phrases "safety-related" and "important to safety."

In determining the reportability of an event or condition that affects a system, it is not necessary to assume an additional random single failure in that system.

The term "safety function" refers to any of the four functions (A through D) listed in these reporting criteria that are required during any plant mode or accident situation as described or relied on in the plant safety analysis report or required by the regulations.

A system must operate long enough to complete its intended safety function as defined in the safety analysis report. Reasonable operator actions to correct minor problems may be considered; however, heroic actions and unusually perceptive diagnoses, particularly during stressful situations, should not be assumed. If a potentially serious human error is made that could have prevented fulfillment of a safety function, but recovery factors resulted in the error being corrected, the error is still reportable.

Both offsite electrical power (transmission lines) and onsite emergency power (usually diesel generators) are considered to be separate functions by GDC 17. If either offsite power or onsite emergency power is unavailable to the plant (i.e., completely lost), it is reportable regardless of whether the other system is available. GDC 17 defines the safety function of each system as providing sufficient capacity and capability, etc., assuming that the other system is not available. Loss of offsite power should be determined at the essential switchgear busses.

As indicated in the Statement of Considerations:

> "The Commission recognizes that the application of this and other paragraphs of this section involves the use of engineering judgment. In this case, a technical judgment must be made whether a failure or operator action that did actually disable one train of a

---

[27]   48 FR 33854, July 28, 1983.

safety system, could have, but did not, affect a redundant train within the ESF system. If so, this would constitute an event that "could have prevented" the fulfillment of a safety function, and, accordingly, must be reported.

- If a component fails by an apparently random mechanism it may or may not be reportable if the functionally redundant component could fail by the same mechanism. Reporting is required if the failure constitutes a condition where there is reasonable doubt that the functionally redundant train or channel would remain operational until it completed its safety function or is repaired. For example, if a pump in one train of an ESF system fails because of improper lubrication, and engineering judgment indicates that there is a reasonable expectation that the functionally redundant pump in the other train, which was also improperly lubricated, would have also failed before it completed its safety function, then the actual failure is reportable and the potential failure of the functionally redundant pump must be discussed in the LER.

For systems that include three or more trains, the failure of two or more trains should be reported if, in the judgment of the licensee, the functional capability of the overall system was jeopardized."[28]

and:

"Finally, the Commission recognizes that the licensee may also use engineering judgment to decide when personnel actions could have prevented fulfillment of a safety function. For example, when an individual improperly operates or maintains a component, he might conceivably have made the same error for all of the functionally redundant components (e.g., if he incorrectly calibrates one bistable amplifier in the Reactor Protection System, he could conceivably incorrectly calibrate all bistable amplifiers). However, for an event to be reportable it is necessary that the actions actually affect or involve components in more than one train or channel of a safety system, and the result of the actions must be undesirable from the perspective of protecting the health and safety of the public. The components can be functionally redundant (e.g, two pumps in different trains) or not functionally redundant (e.g., the operator correctly stops a pump in Train "A" and instead of shutting the pump discharge valve in Train "A," he mistakenly shuts the pump discharge valve in Train "B")."[29]

Any time a system did not or could not have performed its safety function because of a single failure, common-mode failure, or combination of independent failures it is reportable under these criteria. These reporting requirements apply to the system level, rather than the train or component level.

---

[28]   48 FR 33854 and 48 FR 33858, July 26, 1983.

[29]   48 FR 33854 and 48 FR 33858, July 26, 1983.

- Single Failure

    These reporting criteria are not meant to require reporting of a single, independent (i.e., random) component failure that makes only one functionally redundant train inoperative unless it is indicative of a generic problem (i.e., has common-mode failure implications).

    As indicated in Paragraph 50.73(a)(2)(vi) "...individual component failures need not be reported pursuant to this paragraph if redundant equipment in the same system was operable and available to perform the required safety function..."

    The staff considers application of this principle to include cases where one train of a two train system is:

    (1)   failed, or;

    (2)   otherwise incapable of performing its function because of factors such as operator error or design, analysis, fabrication, construction and/or procedural inadequacies; or;

    (3)   in the case of a train which should be running, otherwise not performing its function because of factors such as operator error or design, analysis, fabrication, construction and/or procedural inadequacies; or;

    (4)   otherwise subject to a reasonable expectation of being prevented from fulfilling its safety function.

    The staff believes that the conditions necessary to consider the redundant train operable and available, for this purpose, should include the following:

    (1)   in cases where the redundant train should operate automatically, it is capable of timely and correct automatic operation, or in cases where the redundant train should be operated manually, the operators would detect[30] the need for its operation and initiate such operation, using established procedures for which they are trained, within the needed time frame, without the need for trouble shooting and repair, and;

    (2)   the redundant train is capable of performing its safety function for the duration required, and;

---

[30]   For example, conditions that would indicate a need for operation of the redundant train are regularly monitored and instrumentation used to monitor these conditions is capable and available.

(3)   there is not a reasonable expectation of preventing fulfillment of the safety function by the redundant train.[31]

A single failure that defeats the safety function of a system is reportable even if the design of the system, which allows such a single failure to defeat the function of the system, has been found acceptable.

As discussed in the Statement of Considerations, "there are a limited number of single-train systems that perform safety functions (e.g., the High Pressure Coolant Injection System in BWRs). For such systems, loss of the single train would prevent the fulfillment of the safety function of that system and, therefore, is reportable even though the plant technical specifications may allow such a condition to exist for a limited time."[32]

- **Common-Cause Failures**

    The following conditions are reportable under these criteria:

    - an event or condition that disabled multiple trains of a system because of a single cause

    - an event or condition where one train of a system is disabled; in addition, (1) the underlying cause that disabled one train of a system could have failed a redundant train and (2) there is reasonable expectation that the second train would not complete its safety function if called upon

    - an observed or identified event or condition that alone could have prevented fulfillment of the safety function

- **Multiple equipment inoperability or unavailability**

    Whenever an event or condition exists where the system could have been prevented from fulfilling its safety function because of one or more reasons for equipment

---

[31]   For example, this means that the exclusion from reporting single component failures under this criterion (i.e., Paragraphs 50.72(b)(2)(iii), 50.73(a)(2)(v) and 50.73(a)(2)(vi)) should not apply when there is a reasonable expectation of failure of the redundant train as a result of the same cause. Application of this principle is illustrated in several parts of this section, including: (1) the immediately proceeding quotations from 48 FR 33854 and 48 FR 33858; (2) the immediately following discussion of common cause failures, and; (3) the discussions in Examples 12 and 13. As indicated in the first paragraph of this section, the event should be reported under this criterion if there is a reasonable expectation of preventing fulfillment of the safety function.

[32]   48 FR 33854, July 26, 1983.

inoperability or unavailability, it is reportable under these criteria. This would include cases where one train is disabled and a second train fails a surveillance test.

Reportability of any of the above type failures (single, common-mode, or multiple) under both 10 CFR 50.72 and 50.73 is independent of power or plant mode. It also is independent of whether:

- the system or structure was demanded at the time of discovery

- the system or structure was required to be operable at the time of discovery

- the cause of a potential failure of the system was corrected before an actual demand for the safety function could occur

- other systems or structures were available that could have or did perform the safety function

- the entire system or structure is specified as ESF or safety related

- the problem occurs in a non-safety portion of a system

The following types of events or conditions generally are not reportable under these criteria:

- failures that affect inputs or services to systems that have no safety function (unless it could prevent the performance of a safety function of an adjacent or interfacing system)

- a single defective component that was delivered, but not installed

- removal of a system or part of a system from service as part of a planned evolution for maintenance or surveillance testing when done in accordance with an approved procedure and the plant's TS (unless a condition is discovered that could have prevented the system from performing its function)

- independent failure of a single component (unless it is indicative of a generic problem, it alone could have caused a safety system failure, or it is in a single-train system)

- a procedure error discovered before procedure approval and the error could have resulted in defeating the system function

- a failure of a system used only to warn the operator where no credit is taken for it in any safety analysis and it does not directly control any of the safety functions in the criteria

- a single stuck control rod that alone would not have prevented the fulfillment of a reactor shutdown

- unrelated component failures in several different safety systems

The applicability of these criteria includes those safety systems designed to mitigate the consequences of an accident (e.g., containment isolation, emergency filtration). Hence, minor operational events involving a specific component such as valve packing leaks, which could be considered a lack of control of radioactive material, should not be reported under this paragraph. System leaks or other similar events may, however, be reportable under other sections of the rules.[33]

Examples

Single Train Systems

(1)   Failure of a Single-Train System Preventing Accident Mitigation and Residual Heat Removal

When the licensee was preparing to run a surveillance test, a high-pressure coolant injection (HPCI) flow controller was found inoperable; therefore, the licensee declared the HPCI system inoperable. The plant entered a technical specification requiring that the automatic depressurization, low-pressure coolant injection, core spray, and isolation condenser systems remain operable during the 7-day LCO or the plant had to be shut down. The licensee made an ENS notification within 28 minutes and a followup call after the amplifier on the HPCI flow transmitter was fixed and the HPCI returned to operability.

As discussed above, the loss of a single train safety system such as BWR HPCI is reportable.

(2)   Failure of a Single-Train Non-Safety System

Question:

If RCIC is not a "safety system" in that no credit for its operation is taken in the safety analysis, are failures and unavailability of this system reportable?

Answer:

If the plant's safety analysis considered RCIC as a system needed to remove residual heat (e.g., it is included in the Technical Specifications) then its failure is reportable under this criterion; otherwise, it is not reportable under this section of the rule.

---

[33]   48 FR 33854, July 26, 1983.

(3)   Failure of a Single-Train Environmental System

Question:

There are a number of environmental systems in a plant dealing with such things as low level waste (e.g., gaseous radwaste tanks). Many of these systems are not required to meet the single failure criterion so a single failure results in the loss of function of the system. Are all of these systems covered within the scope of the LER rule?

Answer:

If such systems are required by Technical Specifications to be operational and the system is needed to fulfill one of the safety functions identified in this section of the rule then system level failures are reportable. If the system is not covered by Technical Specifications and is not required to meet the single failure criterion, then failures of the system are not reportable under this criterion.

Loss of Two Trains

(4)   Loss of Onsite Emergency Power by Multiple Equipment Inoperability and Unavailability

During refueling, one emergency diesel generator (EDG) in a two train system was out of service for maintenance. The second EDG was declared inoperable when it failed its surveillance test.

An ENS notification is required and an LER is required. As addressed in the Discussion section above, loss of either the onsite power system or the offsite power system is reportable under this criterion.

(5)   Procedure Error Prevents Reactor Shutdown Function

The unit was in mode 5 (95°F and 0 psig ; before initial criticality) and a post-modification test was in progress on the train A reactor protection system (RPS), when the operator observed that both train A and B source range detectors were disabled. During post-modification testing on train A RPS, instrumentation personnel placed the train B input error inhibit switch in the inhibit position. With both trains' input error inhibit switches in the inhibit position, source range detector voltage was disabled. The input error inhibit switch was immediately returned to the normal position and a caution was added to appropriate plant instructions.

This event is reportable because disabling the source range detectors could have prevented fulfillment of the safety function to shut down the reactor.

(6)   Failure of the Overpressurization Mitigation System

The RCS was overpressurized on two occasions during startup following a refueling outage because the overpressure mitigation system (OMS) failed to operate. The reason that the OMS failed to operate was that one train was out of service for maintenance and a pressure transmitter was isolated and a summator failed in the actuation circuit on the other train.

The event is reportable because the OMS failed to perform its safety function.

(7)   Loss of Salt Water Cooling System and Flooding in Saltwater Pump Bay

During maintenance activities on the south saltwater pump, the licensee was removing the pump internals from the casing when flooding of the pump area occurred. The north saltwater pump was secured to prevent pump damage.

The event is reportable because of the failure of the saltwater cooling system, which is the ultimate heat sink for the facility, to perform its safety function.

(8)   Maintenance Affecting Two Trains

Question:

Some clarification is needed for events or conditions that alone "could have" prevented the fulfillment of a system safety function.

Answer:

"Events or conditions" generally involve operator actions and/or component failures that could have prevented the functioning of a safety system. For example, assume that a surveillance test is run on a standby pump and it seizes. The pump is disassembled and found to contain the wrong lubricant. The redundant pump is disassembled and it also has the same wrong lubricant. Thus, it is reasonable to assume that the second pump would have failed if it had been challenged. However, the second pump and, therefore, the system did not actually fail because the second pump was never challenged. Thus, in this case, because of the use of the wrong lubricant, the system "could have" or "would have" failed.

Loss of One Train

(9)   Oversized Breaker Wiring Lugs

Situation:

During testing of 480 volt safety-related breakers, one breaker would not trip electrically. Investigation revealed that one wire of the pigtail on the trip coil, although still in its lug, was so loose that there was no electrical connection. The loose connection was due to the fact that the pigtail lug was too large (No. 14-16 AWG), whereas the pigtail wire was No. 20 AWG. A No. 18-22 lug is the acceptable industry standard for a No. 20 AWG wire.

Since the trip coils were supplied pre-wired, all safety-related breakers utilizing the trip coil were inspected. All other breakers inspected had No. 14-16 AWG lugs. No lugs were found with loose electrical connections. Nevertheless, all No. 14-16 AWG lugs were replaced with acceptable industry Standard No. 18-22 AWG lugs.

Comment:

The event is reportable because the incompatible pigtails and lugs could have caused one or more safety systems to fail to perform their intended function [50.73(a)(2)(v)].

(10)   Contaminated Hydraulic Fluid Degrades MSIV Operation

Situation:

During a routine shutdown, the operator noted that the #11 MSIV closing time appeared to be excessive. A subsequent test revealed the #11 MSIV shut within the required time, however, the #12 MSIV closing time exceeded the maximum at 7.4 sec. Contamination of the hydraulic fluid in the valve actuation system had caused the system's check valves to stick and delay the transmission of hydraulic pressure to the actuator. Three more filters will be purchased providing supplemental filtering for each MSIV. Finer filters will be used in pump suction filters to remove the fine contaminants. The #12 MSIV was repaired and returned to service. Since the valves were not required for operation at the time of discovery, the safety of the public was not affected.

Comments:

The event is reportable because a single condition could have prevented fulfillment of a safety function [50.73(a)(2)(v)].

The fact that the condition was discovered when the valves were not required for operation does not affect the reportability of the condition.

(11)   Diesel Generator Lube Oil Fire Hazard

Situation:

While performing a routine surveillance test of the emergency diesel generator, a small fire started due to lubricating oil leakage from the exhaust manifold. The manufacturer reviewed the incident and determined that the oil was accumulating in the exhaust manifold due to leakage originating from above the upper pistons of this vertically opposed piston engine. The oil remaining above the upper pistons after shutdown leaked slowly down past the piston rings, into the combustion space, past the lower piston rings, through the exhaust ports, and into the exhaust manifolds. The exhaust manifolds became pressurized during the subsequent startup which forced the oil out through leaks in the exhaust manifold gaskets where it was ignited.

Similar events occurred previously at this plant. In these previous cases, fuel oil accumulated in the exhaust manifold due to extended operation under "no load" conditions. Operation under loaded conditions was therefore required before shutdown in order to burn off any accumulated oil.

Comments:

The event is not reportable if the fire did not pose a threat to the plant (i.e., it only affected a single component) [50.73(a)(2)(x)].

The event would be reportable if it demonstrates a design, procedural, or equipment deficiency that could have prevented the fulfillment of a safety function (i.e., if the redundant diesels are of similar design and, therefore, susceptible to the same problem) [50.73(a)(2)(v)].

(12)   Single Failures

Question:

I notice that loss of relief/safety valve capability is reportable. Does this mean that an LER is required when one valve is inoperative? In addition, suppose your have one pump in a cooling water system (e.g., chilled water) supplying water to both trains of a safety system, but there is another pump in standby; is the loss of the one operating pump reportable?

Answer:

No. Single, independent (i.e., random) component failures are not reportable as LERs if the redundant component in the same system did or would have fulfilled the safety function. In general, however, such failures are reportable to the NPRD system. However, if such failures have generic implications, then an LER is to be submitted.

(See the discussion under the heading "Single Failures" for further discussion of reporting the loss of one train.)

(13)   Generic Setpoint Drift

*   Situation:

With the plant in steady state operation at 2170 MWt and while performing a Main Steam Line Pressure Instrument Functional Test and Calibration, a switch was found to actuate at 853 psig. The Tech Specs limit is 825 +15 psig. The redundant switches were operable. The cause of the occurrence was setpoint drift. The switch was recalibrated and tested successfully per HNP-2-5279, Barksdale Pressure Switch Calibration, and returned to service.

This is a repetitive event as reported in one previous LER. A generic review revealed that these type switches are used on other safety systems and that this type switch is subject to drift. An investigation will continue as to why these switches drift, and if necessary, they will be replaced.

Comments:

The event is not reportable due to the drift of a single pressure switch.

The event is reportable if it is indicative of a generic and/or repetitive problem with this type of switch which is used in several safety systems [50.73(a)(2)(v) or (viii)].

- Question:

Are setpoint drift problems with a particular switch to be reported if they are experienced more than once?

Answer:

The independent failure (e.g., excessive setpoint drift) of a single pressure switch is not reportable unless it alone could have caused a system to fail to fulfill its safety function, or is indicative of a generic problem that could have resulted in the failure of more than one switch and thereby cause one or more systems to fail to fulfill their safety function.

(14) Maintenance Affecting Only One Train

Question:

Suppose the wrong lubricant was installed in one pump, but the pump in the other train was correctly lubricated. Is this reportable?

Answer:

Engineering judgement is required to decide if the lubricant could have been used on the other pump, and, therefore, the system function would have been lost. If the procedure called for testing of the first pump before maintenance was performed on the second pump and testing clearly identified the error, then the error would not be reportable. However, if the procedure called for the wrong lubricant and eventually both pumps would have been improperly lubricated, and the problem was only discovered when the first pump was actually challenged and failed, then the error would be reportable.

<u>Other Conditions</u>

(15)   Conditions Observed While System Out of Service

Question:

Suppose during shutdown we are doing maintenance on both SI pumps, which are not required to be operational. Is this reportable? While shutdown, suppose I identify or observe something that would cause the SI pumps not to be operational at power. Is this reportable?

Answer:

Removing both SI pumps from service to do maintenance is not reportable if the resulting system configuration is not prohibited by the plant's technical specifications. However, if a situation is discovered during maintenance that could have caused both pumps to fail, (e.g., they are both improperly lubricated) then that condition is reportable even though the pumps were not required to be operational at the time that the condition was discovered. As another example, suppose the scram breakers were tested during shutdown conditions, and it was found that for more than one breaker, opening times were in excess of those specified, or that UV trip attachments were inoperative. Such potential generic problems are reportable in an LER.

(16)   Diesel Generator Bearing Problems

During the annual inspection of one standby diesel generator, the lower crankshaft thrust bearing and adjacent main bearing were found wiped on the journal surface. The thrust bearing was also found to have a small crack from the main oil supply line across the journal surface to the thrust surface. Inspection of the second, redundant standby diesel generator annual inspection revealed similar problems. It was judged that extended operation without corrective action could have resulted in bearing failure.

The event is reportable because there was reasonable doubt that both diesels would have remained operable until they completed their safety function if called upon.

(17)   Potential Loss of High Pressure Coolant Injection

During normal refueling leak testing of the upstream containment isolation check valve on the High Pressure Coolant Injection (HPCI) steam exhaust, the disc of the non-containment isolation check valve was found lodged in downstream piping. This might have prevented HPCI from functioning if the disc had blocked the line. The event was caused by fatigue failure of a disc pin.

Following evaluation of the condition, the event was determined to be reportable because the HPCI could have been prevented from performing its safety function if the disc had blocked the line. In addition, the event is reportable if the fatigue failure is indicative of a common-mode failure.

(18)   Defective Component Delivered but not Installed

Question:

How should a plant report a defective component that was delivered, but not installed?

Answer:

A single defective component would not generally be reportable (assuming that the problem has no generic implications). A generic problem or a number of defective components would probably constitute a condition that could have prevented fulfillment of a safety function, and, if so, would be reportable. Engineering judgment is required to determine if the defects could have escaped detection prior to installation and operation. As a minimum, any generic problem may be reported as a voluntary LER. In addition, such a condition may be reportable under 10 CFR Part 21.

(19)   Operator Inaction or Wrong Action

Question:

In some systems used to control the release of radioactivity, a detector controls certain equipment. In other systems, a monitor is present and the operator is required to initiate action under certain conditions. The operator is not "wired" in. Are failures of the operator to act reportable?

Answer:

Yes. The operator may be viewed as a "component" that is an integral, and frequently essential, part of a "system." Thus, if an event or condition meets the criterion specified in 50.73 for reporting, it is to be reported regardless of the initiating cause (i.e., whether an equipment, procedure, or personnel error is involved).

(20)   Results of Analysis

Question:

A number of criteria indicate that they apply to actual situations only and not to potential situations identified as a result of analysis; yet, other criteria address "could have." When do the results of analysis have to be reported?

Answer:

The results need only to be reported if the applicable criterion requires the reporting of conditions that "could have" caused a problem. However, others have a need to know about potential problems that are not reportable; thus, such items may be reported as a voluntary LER.

(21)   System Interactions

Question:

Utilities are not required to analyze for system interactions, yet the rule requires the reporting of events that "could have" happened but did not. Are we to initiate a design activity to determine "could have" system interactions?

Answer:

No. Report system interactions that you find as a result of ongoing routine activities (e.g., the analysis of operating events).

3.3.4 Common-cause Failures of Independent Trains or Channels

10 CFR 50.72	§50.73(a)(2)(vii)
[No corresponding Part 50.72 requirement.]	Licensees shall report: "Any event where a single cause or condition caused at least one independent train or channel to become inoperable in multiple systems or two independent trains or channels to become inoperable in a single system designed to: (A) Shut down the reactor and maintain it in a safe shutdown condition; (B) Remove residual heat; (C) Control the release of radioactive material; or (D) Mitigate the consequences of an accident."

Licensees are required to report a common-cause failure as an LER within 30 days.

Discussion

This criterion requires those events to be reported where a single cause or condition caused independent trains or channels to become inoperable. Common-causes may include such factors as high ambient temperatures, heatup from energization, inadequate preventive maintenance, oil contamination of air systems, incorrect lubrication, use of non-qualified components or manufacturing or design flaws. The event is reportable if the independent trains or channels were inoperable at the same time, regardless of whether or not they were discovered at the same time. (Example (2) below illustrates a case where the second failure was discovered 3 days later than the first.)

An event or failure that results in or involves the failure of independent portions of more than one train or channel in the same or different systems is reportable. For example, if a cause or condition caused components in Train "A" and "B" of a single system to become inoperable, even if additional trains (e.g., Train "C") were still available, the event must be reported. In addition, if the cause or condition caused components in Train "A" of one system and in Train "B" of another system (i.e., train that is assumed in the safety analysis to be independent) to become inoperable, the event must be reported. However, if a cause or condition caused components in Train "A" of one system and Train "A" of another system (i.e., trains that are not assumed in the safety analysis to be independent), the event need not be reported unless it meets one or more of the other reporting criteria.

Trains or channels for reportability purposes are defined as those redundant, independent trains or channels designed to provide protection against single failures. Many engineered safety systems containing active components are designed with at least a two-train system.

Each independent train in a two-train system can normally satisfy all the safety system requirements to safely shut down the plant or satisfy those criteria that have to be met following an accident.

This criterion does not include those cases where one train of a system or a component was removed from service as part of a planned evolution, in accordance with an approved procedure, and in accordance with the plant's technical specifications. For example, if the licensee removes part of a system from service to perform maintenance, and the Technical Specifications permit the resulting configuration, and the system or component is returned to service within the time limit specified in the Technical Specifications, the action need not be reported under this paragraph. However, if, while the train or component is out of service, the licensee identifies a condition that could have prevented the whole system from performing its intended function (e.g, the licensee finds a set of relays that is wired incorrectly), that condition must be reported.

Analysis of events reported under this part of the rule may identify previously unrecognized common-cause (or dependent) failures and system interactions. Such failures can be simultaneous failures that occur because of a single initiating cause (i.e., the single cause or mechanism serves as a common input to the failures); or the failures can be sequential (i.e., cascading failures), such as the case where a single component failure results in the failure of one or more additional components.

Examples

(1) Incorrect Lubrication Degrades Main Steam Isolation Valve Operation

During monthly operability tests, the licensee found that the Unit 2B inboard MSIV did not stroke properly as a result of a solenoid-operated valve (SOV) failure. Both units were shut down from 100-percent power, and the SOVs piloting all 16 MSIVs were inspected. The licensee found that the SOVs on all 16 MSIVs were damaged. The three-way and four-way valves and solenoid pilot valves on all 16 MSIVs had a hardened, sticky substance in their ports and on their O-rings. As a result, motion of all the SOVs was impaired, resulting in instrument air leakage and the inability to operate all of the MSIVs satisfactorily. The licensee also examined unused spares in the warehouse and found that the lubricant had dried out in those valves, leaving a residue. Several of the warehouse spares were bench tested. They were found to be degraded and also leaked. The root cause of the event was use of an incorrect lubricant.

The event is reportable (a) because a single cause or condition caused multiple independent trains of the main steam isolation system (a system designed to control the release of radioactive material and mitigate the consequences of an accident) to become inoperable [§50.73(a)(2)(vii)(C and D)] and (b) because a single condition could have prevented fulfillment of a safety function [§50.73(a)(2)(v)].

(2)   Marine Growth Causing Emergency Service Water To Become Inoperable (Common-Mode Failure Mechanism)

With Unit 1 at 74 percent power and Unit 2 at 100 percent power, ESW pump 1A was declared inoperable because its flow rate was too low to meet acceptance criteria. Three days later, with both units at the same conditions, ESW pump 1C was declared inoperable for the same reason. The ESW pumps provide the source of water to the intake canal during a design-basis accident. In both cases, the cause was marine growth of hydroids and barnacles on the impeller and suction of the pumps. Following maintenance, both pumps passed their performance tests and were placed in service. Pump testing frequency was increased to more closely monitor pump performance.

This event is reportable because a single cause or condition caused two independent trains to become inoperable in a single system designed to mitigate the consequences of an accident [§50.73(a)(2)(vii)(D)].

(3)   Testing Indicated Several Inoperable Snubbers

The licensee found 11 inoperable snubbers during periodic testing. All the snubbers failed to lock up in tension and/or compression. These failures did not render their respective systems inoperable, but rendered trains inoperable. Improper lockup settings and/or excessive seal bypass caused these snubbers to malfunction. These snubbers were designed for low probability seismic events. Numerous previous similar events have been reported by this licensee.

This condition is reportable because the condition indicated a generic common-mode problem that caused numerous multiple independent trains in one or more safety systems to become inoperable. The potential existed for numerous snubbers in several systems to fail following a seismic event rendering several trains inoperable. [§ 50.73(a)(2)(vii)]

(4)   Stuck High-Pressure Injection (HPI) System Check Valves as a Result of Corroded Flappers

The licensee reported that check valves in three of four HPI lines were stuck closed. The unit had been shut down for refueling and maintenance.

A special test of the check valves revealed that three 2½-inch stop check valves remained closed when 130 pounds per square inch (psi) of differential pressure was applied to the valve. An additional test revealed that the valve failed to open when 400 psi of differential pressure (the capacity of the pump) was applied to the valve. Further review showed that the common cause of valve failure was the flappers corroding shut.

The event is reportable because a single cause or condition caused at least two independent trains of the HPI system to become inoperable. This system is designed to remove residual heat and mitigate the consequences of an accident. The condition is therefore reportable under 50.73(a)(2)(vii)(B and D), common cause failure in systems designed to remove residual heat and mitigate accidents.

3.3.5 Airborne or Liquid Effluent Release

§50.72(b)(2)(iv)	§50.73(a)(2)(viii)
Licensees shall report: "(A) Any airborne radioactive release that, when averaged over a time period of 1 hour, results in concentrations in an unrestricted area that exceed 20 times the applicable concentration specified in Appendix B to Part 20, Table 2, Column 1. (B) Any liquid effluent release that, when averaged over a time period of 1 hour, exceeds 20 times the applicable concentration specified in Appendix B to Part 20, Table 2, Column 2, at the point of entry into the receiving waters (i.e., unrestricted area) for all radionuclides except tritium and dissolved noble gases. (Immediate notifications made under this paragraph also satisfy the requirements of §20.2202 of this chapter.)"	Licensees shall report: "(A) Any airborne radioactivity release that, when averaged over a time period of 1 hour, resulted in airborne radionuclide concentrations in an unrestricted area that exceeded 20 times the applicable concentration limits specified in Appendix B to Part 20, Table 2, Column 1. (B) Any liquid effluent release that, when averaged over a time period of 1 hour, exceeds 20 times the applicable concentrations specified in Appendix B to Part 20, Table 2, Column 2, at the point of entry into the receiving waters (i.e., unrestricted area) for all radionuclides except tritium and dissolved noble gases. **§50.73(a)(2)(ix)** Reports submitted to the Commission in accordance with paragraph (a)(2)(viii) of this section also meet the effluent release reporting requirements of §20.2203(a)(3) of this chapter.

If not reported under §50.72(a) or (b)(1), licensees are required to report such airborne or liquid effluent releases as defined in the regulations above to the NRC via the ENS as soon as practical and in all cases within 4 hours of the event. Licensees are required to submit an LER within 30 days.

Discussion

Although similar to 10 CFR 20.403 (20.2202) and 20.405 (20.2203), these criteria place a lower threshold for reporting events at commercial power reactors because the significance of the breakdown of the licensee's program that allowed such a release is the primary concern, rather than the significance of the effect of the actual release.

For a release that takes less than 1 hour, normalize the release to 1 hour (e.g., if the release lasted 15 minutes, divide by 4). For releases that lasted more than 1 hour, use the highest release for any continuous 60-minute period (i.e., comparable to a moving average).

Annual average meteorological data should be used for determining offsite airborne concentrations of radioactivity to maintain consistency with the technical specifications (TS) for reportability thresholds.

The location used as the point of release for calculation purposes should be determined using the expanded definition of an unrestricted area as specified in NUREG-0133 ("Preparation of Radiological Effluent Technical Specifications for Nuclear Power Plants," October 1978) to maintain consistency with the TS.

If estimates determine that the release has exceeded the reporting criterion, an ENS notification is required, followed up by a more precise estimate in the LER. If it is later determined that the release was less than this criterion, the ENS notification may be retracted.

As indicated in Generic Letter 85-19, September 27, 1985, "Reporting Requirements on Primary Coolant Iodine Spikes," primary coolant iodine spike releases need not be reported on a short term basis.

Examples

(1) Unmonitored Release of Contaminated Steam Through Auxiliary Boiler Atmospheric Vent

An unmonitored release of contaminated steam resulted from a combination of a tube leak, improper venting of an auxiliary boiler system, and inadequate procedures. This combination resulted in a release path from a liquid waste concentrator to the atmosphere via the auxiliary boiler system steam drum vent.

Because of rain at the site, the steam release to the atmosphere was condensed and deposited onto plant buildings and yard areas. This contamination was washed via a storm drain into a lake. The release was later confirmed to be 2.6 E-5 µCi/ml of Cs-137 at the point of entry into the receiving water.

An ENS notification is required as a liquid radioactive material release because the unmonitored release exceeded 20 times the applicable concentrations specified in Table 2, Column 2 of Appendix B to 10 CFR Part 20, averaged over 1 hour at the site boundary. An LER is required.

(2) Unplanned Gaseous Release

During routine scheduled maintenance on a pressure actuated valve in the gaseous waste system, an unplanned radioactive release to the environment was detected by a main stack high radiation alarm. The release occurred when an isolation valve, required to be closed on the station tagout sheet, was inadvertently left open. This allowed radioactive gas from the waste gas decay tank to escape through a pressure gage connection that had been opened to vent the system. Operator error was the root cause of this release, with ambiguous valve tag numbers as a contributing factor. The concentration in the unrestricted area, averaged over 1 hour, was estimated by the licensee to be 1 E-5 µCi/ml of Kr-85 and 5 E-6 µCi/ml of Xe-133.

The event was reportable via ENS and LER because the sum of the ratios of the concentration of each airborne radionuclide in the restricted area when averaged over a period of 1 hour, to its respective concentration specified in Table 2, Column 1 of Appendix B to 10 CFR 20, exceeds 20.

3.3.6 Contaminated Person Requiring Transport Offsite

§50.72(b)(2)(v)	10 CFR 50.73
Licensees shall report: "Any event requiring the transport of a radioactively contaminated person to an offsite medical facility for treatment."	[No corresponding Part 50.73 requirement.]

If not reported under §50.72(a) or (b)(1), licensees are required to notify the NRC via the ENS of any such transport as soon as practical and in all cases within 4 hours of the event necessitating the offsite transport.

Discussion

The phrase "radioactively contaminated" refers to either radioactively contaminated clothing and/or person. If there is a potential for contamination (e.g., an initial onsite survey for radioactive contamination is required but has not been completed before transport of the person off site for medical treatment) the licensee should make an ENS notification. See the example.

No LER is required for transporting a radioactively contaminated person to an offsite medical facility for treatment.

Example

(1)   Radioactively Contaminated Person Transported Offsite for Medical Treatment

A contract worker experienced a back injury lifting a tool while working in the reactor containment and was considered potentially contaminated because his back could not be surveyed. Health physics (HP) technicians accompanied the worker to the hospital. The licensee made an ENS notification immediately and an update notification after clothing, but not the individual, was found to be contaminated. The HP technicians returned to the plant with the contaminated protective clothing worn by the worker.

If not reported under §50.72(a)(1) as a declared Unusual Event per the licensee's emergency plan, an ENS notification is required because of the transport of a radioactively contaminated person to an offsite medical facility for treatment.

§50.72(b)(2)(vi)	10 CFR 50.73
Licensees shall report: "Any event or situation, related to the health and safety of the public or on-site personnel, or protection of the environment, for which a news release is planned or notification to other government agencies has been or will be made. Such an event may include an on-site fatality or inadvertent release of radioactively contaminated materials."	[No corresponding Part 50.73 requirement.]

If not reported under §50.72(a) or (b)(1), licensees are required to notify the NRC via the ENS as soon as practical and in all cases within 4 hours of the event, or the decision to prepare a news release, or the decision to notify (or actual notification of) other government agencies.

<u>Discussion</u>

The purpose of this criterion is to ensure the NRC is made aware of issues that will cause heightened public or government concern related to the radiological health and safety of the public or on-site personnel or protection of the environment.

Licensees typically issue press releases or notify local, county, State or Federal agencies on a wide range of topics that are of interest to the general public. The NRC Operations Center does not need to be made aware of every press release made by a licensee. The following clarifications are intended to set a reporting threshold that ensures necessary reporting, while minimizing unnecessary reporting.

Examples of events likely to be reportable under this criterion include

- release of radioactively contaminated tools or equipment to public areas
- unusual or abnormal releases of radioactive effluents
- onsite fatality

Licensees generally do not have to report media and government interactions unless they are related to the radiological health and safety of the public or onsite personnel, or protection of the environment. For example, the NRC does not generally need to be informed under this criterion of:

- minor deviations from sewage or chlorine effluent limits
- minor non-radioactive, onsite chemical spills
- problems with plant stack or water tower aviation lighting

- peaceful demonstrations
- routine reports of effluent releases to other agencies

Press Release

The NRC has an obligation to inform the public about issues within the NRC's purview that affect or raise a concern about the public health and safety. Thus, the NRC needs accurate, detailed information in a timely manner regarding such situations. The NRC should be aware of information that is available for the press or other government agencies.

However, the NRC need not be notified of every press release a licensee issues. The field of NRC interest is narrowed by the phrase "related to the health and safety of the public or onsite personnel, or protection of the environment," in order to exclude administrative matters or those events of no safety significance.

Routine radiation releases are not specifically reportable under this criterion. However, if a release receives media attention, the release is reportable under this criterion.

If possible, licensees should make an ENS notification before issuing a press release because news media representatives will usually contact the NRC public affairs officer shortly after its issuance for verification, explanation, or interpretation of the facts.

Other Government Notifications

For reporting purposes, "other government agencies" refers to local, State or other Federal agencies.

Notifying another Federal agency does not relieve the licensee of the requirement to report to the NRC.

For those plants which provide a State incident response facility with alarm indication coincident with control room alarms, e.g., an effluent radiation monitor alarm, but the actual radiation release is less than the criteria in §50.72(b)(2)(iv), the NRC does not consider these alarm indications as a notification to the State by the licensee. An alarm received at a State facility is in itself not a requirement for notifying the NRC. In so far as this reporting criterion is concerned, the licensee need only notify the NRC when the licensee determines that a reportable release has occurred, or believes a real potential exists for interest on the part of the State, the media, or the public, or a press release is being planned.

Examples

(1)   Onsite Drowning Government Notifications and Press Release

A boy fell into the discharge canal while fishing and failed to resurface. The licensee notified the local sheriff, State Police, U.S. Coast Guard and State emergency agencies. Local news agencies were granted onsite access for coverage of the event. The licensee notified the NRC resident inspector.

As ENS notification is needed because of the fatality on-site, the other government notifications made, and media involvement.

(2)   Licensee Media Inquiries Regarding NRC Findings

As a result of a local newspaper article regarding the findings of an NRC regional inspection of the 10 CFR Part 50, Appendix R, Fire Protection Program, a licensee representative was interviewed on local television and radio stations. The licensee notified State officials and the NRC resident inspector.

The staff does not consider an ENS notification to be needed because the subject of the radio and TV interviews was an NRC inspection.

(3)   County Government Notification

The licensee informed county governments and other organizations of a spurious actuation of several emergency response sirens in a county (for about 5 minutes according to county residents). The licensee also planned to issue a press release.

An ENS notification is needed because county agencies were notified regarding the inadvertent actuation of part of the public notification system. Such an event also would be reportable if the county informs the licensee of the problem because of the concern of the public for their radiological health and safety.

(4)   State Notification of Unscheduled Radiation Release

The licensee reported to the State that they were going to release about 50 curies of gaseous radioactivity to the atmosphere while filling and venting the pressurizer. The licensee then revised their estimate of the release to 153 curies. However, since the licensee had not informed the State within 24 hours of making the release, they had to reclassify the release as "unscheduled" per their agreement with the State. The licensee notified the State and the NRC resident inspector.

An ENS notification is needed because of the State notification of an "unscheduled" release of gaseous radioactivity. The initial notification to the State of the scheduled release does not need an ENS notification because it is considered as a routine notification.

(5)   State Notification of Improper Dumping of Radioactive Waste

The licensee transported two secondary side filters to the city dump as nonradioactive waste but later determined they were radioactive. The dump site was closed and the filters retrieved. The licensee notified the appropriate State agency and the NRC resident inspector.

An ENS notification is needed because of the notification to the State agency of the inadvertent release of radioactively contaminated material off site, which affects the radiological health and safety of the public and environment.

(6)   Reports Regarding Endangered Species

The licensee notified the U.S. Fish & Wildlife Service and a State agency that an endangered species of sea turtle was found in their circulating water structure trash bar. No press release was issued.

An ENS notification is required because of the notification of state and federal agencies regarding the taking of an endangered species. (The NRC has statutory responsibilities regarding protection of endangered species.)

(7)   Routine Agency Notifications

A licensee notified the U.S. Environmental Protection Agency (EPA) that the circulation water temperature rise exceeded the release permit allowable. This event was caused by the unexpected loss of a circulating water pump while operating at 92-percent power. The licensee reduced power to 73 percent so that the circulating water temperature would decrease to within the allowable limits until the pump could be repaired.

A licensee notified the Federal Aviation Agency that it removed part of its auxiliary boiler stack aviation lighting from service to replace a faulty relay.

A licensee notified the State, EPA, U.S. Coast Guard and Department of Transportation that 5 gallons of diesel fuel oil had spilled onto gravel-covered ground inside the protected area. The spill was cleaned up by removing the gravel and dirt.

The staff does not consider an ENS notification to be needed because these events are routine and have little significance.

3.3.8 Spent Fuel Storage Cask Notifications

§ 50.72(b) (2) (vii)	10 CFR 50.73
Licensees shall report: "Any instance of: (A) A defect in any spent fuel storage cask structure, system, or component which is important to safety; or (B) A significant reduction in the effectiveness of any spent fuel storage cask confinement system during use of the storage cask under a general license issued under §72.210 of this chapter. A followup written report is required by §72.216(b) of this chapter including a description of the means employed to repair any defects or damage and prevent recurrence, using instructions in §72.4, within 30 days of the report submitted in paragraph (a). A copy of the written report must be sent to the administrator of the appropriate Nuclear Regulatory Commission regional office shown in Appendix D to part 20 of this chapter."	[No corresponding Part 50.73 requirement.]

If not reported under §50.72(a) or (b)(1), licensees are required to report any such instances to the NRC via the ENS as soon as practical, and in all cases within 4 hours. A followup written report is required by §72.216(b) within 30 days.

Discussion

This information is necessary to inform the NRC of potential hazards to the public health and safety. The definition of "defect" in 10 CFR 21.3 is compatible with the intent of this reporting requirement. If the defect is evaluated and reported via this reporting criterion of §50.72, then as indicated in §21.2(c), the evaluation and notification obligations of 10 CFR Part 21 are met. (See Section 5.1.9 for further discussion of Part 21 reporting.)

3.4 Followup Notification

This section addresses §50.72(c), "Followup Notification." These notifications are in addition to making the required initial telephone notifications under §50.72(a) or (b). Reporting under this paragraph is intended to provide the NRC with timely notification when an event becomes more serious or additional information or new analysis clarify an event. The paragraph also authorizes the NRC to maintain a continuous communications channel for acquiring necessary followup information.

§50.72(c)	10 CFR 50.73
"*Followup Notification*. With respect to the telephone notifications made under paragraphs (a) and (b) of this section, in addition to making the required initial notification, each licensee shall, during the course of the event: (1) *Immediately report* 　(i) any further degradation in the level of safety of the plant or other worsening plant conditions, including those that require the declaration of any of the Emergency Classes, if such a declaration has not been previously made, or 　(ii) any change from one Emergency Class to another, or 　(iii) a termination of the Emergency Class. (2) *Immediately report* 　(i) the results of ensuing evaluations or assessments of plant conditions, 　(ii) the effectiveness of response or protective measures taken, and 　(iii) information related to plant behavior that is not understood. (3) Maintain an open, continuous communication channel with the NRC Operations Center upon request by the NRC."	[No corresponding Part 50.73 requirement.]

Discussion

These criteria are intended to provide the NRC with timely notification when an event becomes more serious or additional information or new analyses clarify an event. They also permit the NRC to maintain a continuous communications channel because of the need for continuing followup information or because of telecommunications problems.

With regard to the open, continuous communications channel, licensees have a responsibility to provide enough on-shift personnel, knowledgeable about plant operations and emergency plan implementation, to enable timely, accurate, and reliable reporting of operating events without interfering with plant operation as discussed in the Statement of Considerations for the rule and Information Notice 85-80, "Timely Declaration of an Emergency Class, Implementation of an Emergency Plan, and Emergency Notifications."

4 EMERGENCY NOTIFICATION SYSTEM REPORTING

This section describes the ENS referenced in 10 CFR 50.72 and provides general and specific guidelines for ENS reporting.

4.1 Emergency Notification System

The NRC Operations Center is the nucleus of the ENS and has the capability to handle emergency communication needs. The NRC's response to both emergencies and non-emergencies is coordinated in this communication center. The key NRC emergency communications personnel, the emergency officer (EO), regional duty officer (RDO), and the headquarters operations officer (HOO), are trained to notify appropriate NRC personnel and to focus appropriate NRC management attention on any significant event.

(1)   ENS Telephones

Each commercial nuclear power reactor facility has ENS telephones funded by the NRC. These telephones are located in each licensee's control room, technical support center (TSC), and emergency operations facility (EOF). A separate ENS line is installed at EOF's which are not onsite. The ENS is part of the Federal Telecommunications System (FTS). This FTS ENS replaces the dedicated ENS ringdown telephones used previously to provide a reliable communications pathway for event reporting.

(2)   Health Physics Network Telephones

The health physics network (HPN) is designed to provide health physics and environmental information to the NRC Operations Center in the event of an ongoing emergency.

These telephones are installed in each licensee's TSC and EOF and, like the ENS, they are now part of the FTS.

(3)   Tape Recording

The NRC tape-records all conversations with the NRC Operations Center. The tape is saved for a month in case there is a public or private inquiry.

(4)   Facsimile Transmission (Fax)

Licensees occasionally fax an event notification into the NRC Operations Center on a commercial telephone line before making an ENS notification. However, §50.72

requires that licensees notify the NRC Operations Center via the ENS; therefore, licensees also must make an ENS notification.

4.2 General ENS Notification

4.2.1 Timeliness

The required timing for ENS reporting is spelled out in §§50.72(a)(3), (b)(1), (b)(2), (c)(1), (c)(2), and in the Statements of Considerations, as "immediate" and "as soon as practical and in all cases within one (or four) hour(s)" of the occurrence of an event (depending on its significance). The intent is to require licensees to make and act on reportability decisions in a timely manner so that ENS notifications are made to the NRC as soon as practical, keeping in mind the safety of the plant. See Section 2.11 for further discussion of reporting timeliness.

4.2.2 Voluntary Notifications

Licensees may make voluntary or courtesy ENS notifications about events or conditions in which the NRC may be interested. The NRC responds to any voluntary notification of an event or condition as its safety significance warrants, regardless of the licensee's classification of the reporting requirement. If it is determined later that the event is reportable, the licensee can change the ENS notification to a required notification under the appropriate 10 CFR 50.72 reporting criterion.

4.2.3 ENS Notification Retraction

If a licensee makes a 10 CFR 50.72 ENS notification and later determines that the event or condition was not reportable, the licensee should call the NRC Operations Center on the ENS telephone to retract the notification and explain the rationale for that decision. There is no set time limit for ENS telephone retractions. However, since most retractions occur following completion of engineering and/or management review, it is expected that retractions would occur shortly after such review. See section 2.10 for further discussion of retractions.

4.2.4 ENS Event Notification Worksheet (NRC Form 361)

The ENS Event Notification Worksheet (NRC Form 361) is an attachment to Information Notice 89-89, dated December 26, 1989, subject: Event Notification Worksheets. The worksheet provides the usual order of questions and discussion for easier communication and its use often enables a licensee to prepare answers for a more clear and complete notification. A clear ENS notification helps the HOO to understand the safety significance of the event.
Licensees may obtain an event number and notification time from the HOO when the ENS notification is made. If an LER is required, the licensee may include this information in the LER to provide a cross reference to the ENS notification, making the event easier to trace.

Licensees should use proper names for systems and components, as well as their alphanumeric identifications during ENS notifications. Licensees should avoid using local jargon for plant components, areas, operations, and the like so that the HOO can quickly understand the situation and have fewer questions. In addition, others not familiar with the plant can more readily understand the situation.

4.3 Typical ENS Reporting Issues

At the time of an ENS notification, the NRC must independently assess the status of the reactor to determine if it is in a safe condition and expected to remain so. The HOO needs to understand the safety significance of each event to brief NRC management or initiate an NRC response. The HOO will be primarily concerned about the safety significance of the event, the current condition of the plant, and the possible near-term effects the event could have on plant safety. The HOO will attempt to obtain as complete a description as is available at the time of the notification of the event or condition, its causes, and its effects. Depending upon the licensee's description of the event, the HOO may be concerned about other related issues. The questions that the licensees typically may be asked to discuss do not represent a requirement for reporting. These questions are of a nature to allow the HOO information to more fully understand the event and its safety significance and are not meant in any way to distract the licensee from more important issues.

The licensee's first responsibility during a transient is to stabilize the plant and keep it safe. However, licensees should not delay declaring an emergency class when conditions warrant because delaying the declaration can defeat the appropriate response to an emergency. Because of the safety significance of a declared emergency, time is of the essence. The NRC needs to become aware of the situation as soon as practical to activate the NRC Operations Center and the appropriate NRC regional incident response center, as necessary, and to notify other Federal agencies.

The effectiveness of the NRC response during an event depends largely on complete and accurate reporting from the licensee. During an emergency, the appropriate regional incident response center and the NRC Operations Center become focal points for NRC action. Licensee actions during an emergency are monitored by the NRC to ensure that appropriate action is being taken to protect the health and safety of the public. When required, the NRC supports the licensee with technical analysis and coordinates logistics support. The NRC keeps other Federal agencies informed of the status of an incident and provides information to the media. In addition, the NRC assesses and, if necessary, confirms the appropriateness of actions recommended by the licensee to local and State authorities.

Information Notice 85-80, "Timely Declaration of an Emergency Class, Implementation of an Emergency Plan, and Emergency Notification," dated October 15, 1985, indicates that it is the licensee's responsibility to ensure that adequate personnel, knowledge about plant conditions and emergency plan implementing procedures, are available on shift to assist the shift supervisor to classify an emergency and activate the emergency plan, including making appropriate notifications, without interfering with plant operation. When 10 CFR 50.72 was published, the NRC made clear its intent in the Statements of Consideration that notifications on the ENS to the NRC Operations Center should be made by those knowledgeable of the event. If the description of any emergency is to be sufficiently accurate and timely to meet the intent of the NRC's regulations, the personnel responsible for notification must be properly trained and sufficiently knowledgeable of the event to report it correctly. The NRC did not intend that notifications made pursuant to 10 CFR 50.72 would be made by those who did not understand the event that they are reporting.

ENS reportability evaluations should be concluded and the ENS notification made as soon as practical and in all cases within 1 hour or 4 hours to meet 10 CFR 50.72. The Statement of Considerations noted that the 1-hour deadline is necessary if the NRC is to fulfill its responsibilities during and following the most serious events occurring at operating nuclear power plants without interfering with the operator's ability to deal with an accident or transient in the first few critical minutes (48 FR 39041, August 29, 1983).

5 LICENSEE EVENT REPORTS

This section discusses the guidelines for preparing and submitting LERs. Section 5.1 addresses administrative requirements and provides guidelines for submittal; Section 5.2 addresses the requirements and guidelines for the LER content. Portions of the rule are quoted, followed by explanation, if necessary. A copy of the required LER form (NRC Form 366), LER Text Continuation form (NRC Form 366A), and LER Failure Continuation form (NRC Form 366B), are shown at the end of this section. The use of LER information and the review programs associated with LERs are explained in Appendix C.

5.1 LER Reporting Guidelines

This section addresses administrative requirements and provides guidelines for submittal. Topics addressed include submission of reports, forwarding letters, cancellation of LERs, report legibility, reporting exemptions, reports other than LERs that use LER forms, supplemental information, revised reports, and general instructions for completing LER forms.

5.1.1 Submission of LERs

---

§50.73(d)

"Licensee Event Reports must be prepared on Form NRC 366 and submitted within 30 days of discovery of a reportable event or situation to the U.S. Nuclear Regulatory Commission, as specified in §50.4."

An LER is to be submitted (mailed) within 30 days of the discovery date. If a 30-day period ends on a Saturday, Sunday, or holiday, reports submitted on the first working day following the end of the 30 days are acceptable. If a licensee knows that a report will be late or needs an additional day or so to complete the report, the situation should be discussed with the appropriate NRC regional office. See Section 2.11 for further discussion of discovery date.

5.1.2 LER Forwarding Letter and Cancellations

The cover letter forwarding an LER to the NRC should be signed by a responsible official. There is no prescribed format for the letter. The date the letter is issued and the report date should be the same. Licensees are encouraged to include the NRC resident inspector and the Institute of Nuclear Power Operations (INPO) in their distribution. Multiple LERs can be forwarded by one forwarding letter.

Cancellations of LERs submitted should be made by letter. The bases for the cancellation should be explained so that the staff can understand and review the reasons supporting the determination. The notice of cancellation will be filed and stored with the LER and acknowledgement made in various automated data systems.

5.1.3 Report Legibility

---

§50.73(e)

"The reports and copies that licensees are required to submit to the Commission under the provisions of this section must be of sufficient quality to permit legible reproduction and micrographic processing."

No further explanation is necessary.

5.1.4 Exemptions

---

§50.73(f)

"Upon written request from a licensee including adequate justification or at the initiation of the NRC staff, the NRC Executive Director for Operations may, by a letter to the licensee, grant exemptions to the reporting requirements under this section."

Exemptions may be plant specific or generic. However, one of the goals of the LER rule is a consistent set of reporting requirements that apply to all plants. To minimize inconsistencies in the reporting, plant-specific exemptions will not be issued unless justified by unique plant conditions.

5.1.5 Voluntary LERs

Indicate information-type LERS (i.e., voluntary LERs) by checking the "Other" block in Item 11 of the LER form and type "Voluntary Report" in the space immediately below the block. Also give a sequential LER number to the voluntary report as noted in Section 5.2.4(5). Because not all requirements of §50.73(b), "Contents," may pertain to some voluntary reports, licensees should develop the content of such reports to best present the information associated with the situation being reported.

See Section 2.9 for additional discussion of voluntary LERs.

5.1.6 Supplemental Information and Revised LERs

---

§50.73(c)

"The Commission may require the licensee to submit specific additional information beyond that required by paragraph (b) of this section if the Commission finds that supplemental material is necessary for complete understanding of any unusually complex or significant event. These requests for supplemental information will be made in writing and the licensee shall submit, as specified in §50.4, the requested information as a supplement to the initial LER."

This provision authorizes the NRC staff to require the licensee to submit specific supplemental information.

If an LER is incomplete at the time of original submittal or if it contains significant incorrect information of a technical nature, the licensee should use a revised report to provide the additional information or to correct technical errors discovered in the LER. Identify the revision to the original LER in the LER number as described in Section 5.2.4(5).

The revision should be complete and should not contain only supplementary or revised information to the previous LER because the revised LER will replace the previous report in the computer file. In addition, indicate in the text on the LER form the revised or supplementary information by placing a vertical line in the margin.

If an LER mentions that an engineering study was being conducted, report the results of the study in a revised LER only if it would significantly change the reader's perception of the course, significance, implications, or consequences of the event or if it results in substantial changes in the corrective action planned by the licensee.

Use revisions only to provide additional or corrected information about a reported event. Do not use a revision to report subsequent failures of the same or like component, except as permitted in 10 CFR 50.73. Some licensees have incorrectly used revisions to report new events that were discovered months after the original event because they were loosely related to the original event. These revisions had different event dates and discussed new, although similar, events. Report events of this type as new LERs and not as revisions to previous LERs.

If a criterion for reportability was checked in Item 11 of NRC Form 366 and later it was determined that other requirements also pertain, a revised LER should be submitted. When a voluntary LER is submitted and later it was determined that the event was required to be reported, submit a revised LER to identify this fact.

5.1.7 Special Reports

There are a number of requirements in various sections of the technical specifications that require reporting of operating experience that is not covered by 10 CFR 50.73. If LER forms are used to submit special reports, check the "Other" block in item 11 of the form and type "Special Report" in the space immediately below the block. The provisions of §50.73(b) may not be applicable or appropriate in a special report. Develop the content of the report to best present the information associated with the situation being reported. In addition, if the LER form is used to submit a special report, use a report number from the sequence used for LERs.

If an event is reportable both under 10 CFR 50.73 and as a special report, check the block in Item 11 for the applicable section of 50.73 as well as the "Other" block for a special report. The content of the report should depend on the reportable situation.

5.1.8   Appendix J Reports (Containment Leak Rate Test Reports)

A licensee must perform containment integrated and local leak rate testing and report the results as required by Appendix J to 10 CFR Part 50. When the leak rate test identifies a 10 CFR 50.73 reportable situation (see Section 3.2.4 or 3.3.1 of this report), submit an LER and include the results in an Appendix J report by reference, if desired. The LER should address only the reportable situation, not the entire leak rate test.

5.1.9   10 CFR Part 21 Reports

10 CFR Part 21, "Reporting of Defects and Noncompliance," as amended during 1991, encourages licensees of operating nuclear power plants to reduce duplicate evaluation and reporting effort by evaluating deviations in basic components under the 10 CFR 50.72, 50.73, and 73.71 reporting criteria. As indicated in 10 CFR 21.2(c) "For persons licensed to operate a nuclear power plant under Part 50 of this chapter, evaluation of potential defects and appropriate reporting of defects under §§ 50.72, 50.73. or § 73.71 of this chapter satisfies each person's evaluation, notification, and reporting obligation to report defects under this part ...." As discussed in the Statement of Considerations for 10 CFR 21[34], the only case where a defect in a basic component of an operating reactor might be reportable under Part 21, but not under §§ 50.72, 50.73, or 73.71 would involve Part(s) on the shelf. This type of defect, if it does not represent a condition reportable under §§ 50.72 or 50.73, might still represent a condition reportable under 10 CFR Part 21.

For an LER, if the defect meets one of the criteria of 10 CFR 50.73, check the applicable paragraph in Item 11 of NRC Form 366 (LER Form). Licensees are also encouraged to check the "Other" block and indicate "Part 21" in the space immediately below if the defect in a basic component could create a substantial safety hazard. The wording in Item 16 ("Abstract") and Item 17 ("Text") should state that the report constitutes a Part 21 notification. If the defect is applicable to other facilities at a multi-unit site, a single LER may be used by indicating the other involved facilities in Item 8 on the LER Form.

---

[34]   56 FR 36081, July 31, 1991.

5.1.10 Section 73.71 Reports

Submit events or conditions that are reportable under 10 CFR 73.71 using the LER forms with the appropriate blocks in Item 11 checked. If the report contains safeguards information as defined in 10 CFR 73.21, the LER forms may still be used, but should be appropriately marked in accordance with 10 CFR 73.21. Include safeguards and security information only in the narrative and not in the abstract. In addition, the text should clearly indicate the information that is safeguards or security information. Finally, the requirements of §73.21(g) must be met when transmitting safeguards information. For additional guidelines on 10 CFR 73.71 reporting, see Regulatory Guide 5.62, Revision 1, "Reporting of Safeguards Events," November 1987; NUREG-1304, "Reporting of Safeguards Events," February 1988; and Generic Letter 91-03, "Reporting of Safeguards Events," March 6, 1991.

If the LER contains proprietary information, mark it appropriately in Item 17 (text) on of the LER form. Include proprietary information only in the narrative and not in the abstract. In addition, indicate clearly in the narrative the information that is proprietary. Finally, the requirements of §2.790(b) must be met when transmitting proprietary information.

5.1.11 Availability of LER Forms

The NRC will provide LER forms (i.e., NRC Forms 366, 366A, and 366B) free of charge. Copies may be obtained by writing to the NRC Information and Records Management Branch, Office of the Chief Information Officer, US Nuclear Regulatory Commission, Washington, DC 20555. Electronic versions are also available. Licensees are encouraged to use these forms to assist the NRC's processing of the reports.

5.2 LER Content Requirements and Preparation Guidance

Licensees are required to prepare an LER for those events or conditions that meet one or more of the criteria contained in §50.73(a). Paragraph 50.73(b), "Contents," specifies the information that an LER should contain with further explanation when appropriate.

In 1986, the NRC decided to use an optical character reader (OCR) to read LER abstracts into NRC LER data bases (IE Information Notice No. 86-08, "Licensee Event Report (LER) Format Modification," February 3, 1986). At that time, licensees were asked to help reduce the number of errors incurred by the OCR as a result of incompatible print styles by using OCR-compatible typography for preparing LERs. Therefore, certain limitations have been placed on the use of type styles and symbols for the abstract and text of the LERs. These limitations are listed below. (See the Information Notice for details.)

Type Styles:

- Prestige Elite (12 pitch)
- Letter Gothic (12 pitch)
- OCR-B (12 pitch)
- Courier 12 (12 pitch)
- Elite (12 pitch)

- Courier 10 (10 pitch)
- OCR-A (10 pitch)
- Prestige Pica (10 pitch)
- Prestige Pica (10 pitch)

In addition, the following proportional space type-styles can be read: Madeleine, Cubic, Bold, and Title.

It is suggested that output be on typewriter or formed character (letter-quality or near letter-quality) printer (e.g., daisy wheel, laser, ink-jet).

It is suggested that output have an uneven right margin (i.e., we suggest that you not right justify output).

It is suggested that text of the abstract be kept at least ½-inch inside the border on all sides of the area designated for the abstract on the LER form. Text running into the border can interfere with scanning the document.

It is suggested that you do not use underscore, do not use bold print, do not use Italic print style, do not end any lines with a hyphen and do not use paragraph indents. Instead, print copy single space with a blank line between paragraphs.

Limitations on the use of symbols in the textual areas:

- Spell out the word "degree."

- Use </= for "less than or equal to."

- Use >/= for "greater than or equal to."

- Use +/- for "plus or minus."

- Spell out all Greek letters.

Do not use exponents. A number should either be expressed as a decimal, spelled out, or preferably designated in terms of "E" (E field format). For example, 4.2×10^{-6} could be expressed as 4.2E-6, 0.0000042, or 4.2 x 10(-6).

Define all abbreviations and acronyms in both the text and the abstract and explain all component designators the first time they are used (e.g., the emergency service water pump 1-SW-P-1A).

5.2.1 Narrative Description or Text (NRC Form 366A, Item 17)

(1) General

> **§50.73(b)(2)(i)**
>
> The LER shall contain: "A clear, specific, narrative description of what occurred so that knowledgeable readers conversant with the design of commercial nuclear power plants, but not familiar with the details of a particular plant, can understand the complete event."

There is no prescribed format for the LER text; write the narrative in a format that most clearly describes the event. Although §50.73(b) defines the information that should be included, it is not intended as an outline of the text format. After the narrative is written, however, review the appropriate sections of §50.73(b) to make sure that applicable subjects have been adequately addressed. It is helpful to use headings to improve readability. For example, some LERs employ major headings such as event description, safety consequences, corrective actions, and previous similar events and subheadings such as initial conditions, dates and times, event classification, systems status, event or condition causes, failure modes, method of discovery, component information, immediate corrective actions, and actions to prevent recurrence.

Explain exactly what happened during the entire event or condition, including how systems, components, and operating personnel performed. Do not cover specific hardware problems in excessive detail. Describe unique characteristics of a plant as well as other characteristics that influenced the event (favorably or unfavorably). Avoid using plant-unique terms and abbreviations, or, as a minimum, clearly define them. The audience for LERs is large and does not necessarily know the details of each plant.

Include the root causes, the plant status before the event, and the sequence of occurrences. Describe the event from the perspective of the operator (i.e., what the operator saw, did, perceived, understood, or misunderstood). Specific information that should be included, as appropriate, is described in paragraphs 50.73(b)(2)(ii), (b)(3), (b)(4), and (b)(5) of the rule and separately in the following sections.

If several engineered safety feature (ESF) systems actuate during an event, describe all aspects of the complete event, including all actuations sequentially, and those aspects that by themselves would not be reportable. For example, if a random single component failure (generally not reportable) occurs following a reactor scram (reportable), describe the component failure in the narrative of the LER for the reactor scram. There is no need to provide redundant information or unimportant details, but It is necessary to discuss the performance and status of ESF equipment important for defining and understanding what happened and for determining the potential implications of the event.

Paraphrase pertinent sections of the latest submitted safety analysis report (SAR) rather than referencing them because not all organizations or individuals have access to SARs. Extensive cross-referencing would be excessively time consuming considering the large number of LERs and large number of reviewers that read each LER. Ensure that each applicable component's safety-significant effect on the event or condition is clearly and completely described.

Do not use statements such as "this event is not significant with respect to the health and safety of the public" without explaining the basis for the conclusion.

---

§50.73(b)(2)(ii)(A)

The narrative description must include: "Plant operating conditions before the event."

Describe the plant operating conditions such as power level or, if not at power, describe mode, temperature, and pressure that existed before the event.

---

§50.73(b)(2)(ii)(B)

The narrative description must include: "Status of structures, components, or systems that were inoperable at the start of the event and that contributed to the event."

If there were no structures, systems, or components that were inoperable at the start of the event and contributed to the event, so state. Otherwise, identify SSCs that were inoperable and contributed to the initiation or limited the mitigation of the event. This should include alternative mitigating SSCs that are a part of normal or emergency operating procedures that were or could have been used to mitigate, reduce the consequences of, or limit the safety implications of the event. Include the impact of support systems on mitigating systems that could have been used.

---

§50.73(b)(2)(ii)(C)

The narrative description must include: "Dates and approximate times of occurrences."

For a transient or ESF actuation event, the event date and time are the date and time the event actually occurred. If the event is a discovered condition for which the occurrence date is not known, the event date should be specified as the discovery date. However, a discussion of the best estimate of the event date and its basis should be provided in the narrative. For example, if a design deficiency was identified on March 27, 1997 that involved a component installed during refueling in the spring of 1986, and only the discovery date is known with certainty, the event date should be specified as the discovery date. A discussion should be provided that describes, based on the best information available, the most likely time that the design flaw was introduced into the component (e.g, by manufacturer or by plant engineering prior to procurement). The length of time that the component was in service should also be provided (i.e., when it was installed).

Discuss both the discovery date and the event date if they differ. If an LER is not submitted within 30 days from the event date, explain the relationship between the event date, discovery date, and report date in the narrative. See Section 2.11 for further discussion of discovery date.

Give dates and approximate times for all major occurrences discussed in the LER (e.g., discoveries; immediate corrective actions; systems, components, or trains declared inoperable or operable; reactor trip; actuation and termination of equipment operation; and stable conditions achieved). In particular, for standby pumps and emergency generators, indicate the length of time of operation and any intermittent periods of shutdown or inoperability during the event. Include an estimate of the time and date of failure of systems, components, or trains if different from the time and date of discovery. A chronology may be used to clarify the timing of personnel and equipment actions.

For equipment that was inoperable at the start of the event, provide an estimate of the time the equipment became inoperable and the last time the equipment was known to be operable. Indicate the basis for this conclusion (e.g., a test was successfully run or the equipment was operating). For equipment that failed, provide the failure time and the last time the equipment was known to be operable. Also provide the basis for the last time known operable.

Components such as valves and snubbers may be tested over a period of several weeks. During this period, a number of inoperable similar components may be discovered.[35] In such cases, similar failures that are reportable and that are discovered during a single test program within the 30 days of discovery of the first failure may be reported as one LER. For similar failures that are reportable under Section 50.73 criteria and that are discovered during a single test program or activity, report all failures that occurred within the first 30 days of discovery of the first failure on one LER. However, the 30-day clock starts when the first reportable event is discovered. State in the LER text (and code the information in Items 14 and 15) that a supplement to the LER will be submitted when the test is completed. Submit a revision to the original LER when the test is completed. Include all the failures, including those reported in the original LER, in the revised LER (i.e., the revised LER should stand alone).

(3)   Failures and Errors

---

§50.73(b)(2)(ii)(D)

The narrative description must include: "The cause of each component or system failure or personnel error, if known."

Include the root cause(s) identified for each component or system failure (or fault) or personnel error. Contributing factors may be discussed as appropriate. For example, a valve stem

---

[35]   Note that inoperable similar components might indicate common cause failures of independent trains or channels, which are reportable under §50.73(a)(2)(vii); see Section 3.3.4 for further discussion.

breaking could have been caused by a limit switch that had been improperly adjusted during maintenance; in this case, the root cause might be determined to be personnel error and additional discussion could focus on the limit switch adjustment. If the personnel error is determined to have been caused by deficient procedures or inadequate personnel training, this should be explained.

If the cause of a failure cannot be readily determined and the investigation is continuing, the LER should indicate what additional investigation is planned. A supplemental LER should be submitted following the additional investigation if substantial information is identified that would significantly change a reader's perception of the course or consequences of the event, or if there are substantial changes in the corrective actions planned by the licensee.

---

§50.73(b)(2)(ii)(E)

The narrative description must include: "The failure mode, mechanism, and effect of each failed component, if known."

Include the failure mode, mechanism (immediate cause), and effect of each failed component in the narrative. The effect of the failure on safety systems and functions should be fully described. Identify the specific piece part that failed and the specific trains and systems rendered inoperable or degraded. Identify all dependent systems rendered inoperable or degraded. Indicate whether redundant trains were operable and available.

If the equipment is degraded, but not failed, describe the degradation and its effects and indicate why the equipment would still perform its intended function.

---

§50.73(b)(2)(ii)(F)

The narrative description must include: "The Energy Industry Identification System component function identifier and system name of each component or system referred to in the LER.

(1) The Energy Industry Identification System is defined in: IEEE Std 803-1983 (May 16, 1983) Recommended Practice for Unique Identification in Power Plants and Related Facilities--Principles and Definitions.

(2) IEEE Std 803-1983 has been approved for incorporation by reference by the Director of the Federal Register.

A notice of any changes made to the material incorporated by reference will be published in the *Federal Register*. Copies may be obtained from the Institute of Electrical and Electronics Engineers, 345 East 47th Street, New York, NY 10017. IEEE Std 803-1983 is available for inspection at the NRC's Technical Library, which is located in the Phillips

(Continued on next page)

---

§50.73(b)(2)(ii)(F) (Continued)

Building, 7920 Norfolk Avenue, Bethesda, Maryland; and at the Office of the Federal Register, 1100 L Street, NW, Washington, DC."

Note: The NRC library is now located in the Two White Flint North building, 11545 Rockville Pike, Rockville, Maryland.

The system name may be either the full name (e.g., reactor coolant system) or the two-letter system code (such as AB for the reactor coolant system). However, when the name is long (e.g., low-pressure coolant injection system), the system code (e.g., BO) should be used. If the full names are used, The Energy Industry Identification System (EIIS) component function identifier and/or system identifier (i.e., the two letter code) should be included in parentheses following the first reference to a component or system in the narrative. The component function identifiers and system identifiers need not be repeated with each subsequent reference to the same component or system.

Whenever an uncertainty arises concerning the interpretation of a system boundary, for those systems included in the nuclear plant reliability data system (NPRDS) reportable scope, the boundary should be defined consistent with the comparable system descriptions and interpretations contained in the NPRDS Reportable System and Component Scope Manual. If a component within the scope of the Equipment Performance and Information Exchange (EPIX) System is involved, the system and train designation should be consistent with the EIIS used in EPIX.

---

§50.73(b)(2)(ii)(G)

The narrative description must include the following specific information as appropriate for the particular event: "For failures of components with multiple functions, include a list of systems or secondary functions that were also affected."

No further explanation is necessary.

---

§50.73(b)(2)(ii)(H)

The narrative description must include: "For failure that rendered a train of a safety system inoperable, an estimate of the elapsed time from the discovery of the failure until the train was returned to service."

No further explanation is necessary.

---
§50.73(b)(2)(ii)(I)

The narrative description must include: "The method of discovery of each component or system failure or procedural error."

Explain how each component failure, system failure, personnel error, or procedural deficiency was discovered. Examples include reviewing surveillance procedures or results of surveillance tests, pre-startup valve lineup check, performing quarterly maintenance, plant walkdown, etc.

---
§50.73(b)(2)(ii)(J)

The narrative description must include the following specific information as appropriate for the particular event:

"(1) Operator actions that affected the course of the event, including operator errors, procedural deficiencies, or both, that contributed to the event.

(2) For each personnel error, the licensee shall discuss:

(i) Whether the error was a cognitive error (e.g., failure to recognize the actual plant condition, failure to realize which systems should be functioning, failure to recognize the true nature of the event) or a procedural error;

(ii) Whether the error was contrary to an approved procedure, was a direct result of an error in an approved procedure, or was associated with an activity or task that was not covered by an approved procedure;

(iii) Any unusual characteristics of the work location (e.g., heat, noise) that directly contributed to the error; and

(iv) The type of personnel involved (i.e., contractor personnel, utility-licensed operator, utility non-licensed operator, other utility personnel)."

Human performance often influences the outcome of nuclear power plant events. Human error is known to contribute to more than half of the LERs. The LER rule identifies the types of reactor events and problems that are believed to be significant and useful to the NRC in its effort to identify and resolve threats to public safety. It is designed to provide the information necessary for engineering studies of operational anomalies and trends and patterns analysis of operational occurrences including human performance.

Generally, the criteria of Section 50.73(b)(2)(i) require a clear, specific narrative so that knowledgeable readers can understand the complete event. Further, the criteria of Section 50.73(b)(2)(ii)(J) require a description of (1) operator actions that affected the course of the event and (2) for each personnel error, additional specific information as detailed in the rule. In order to support an understanding of human performance issues related to the event, address the factors discussed below to the extent they apply. For example, if an operator error that affected the course of the event was due to a procedural problem, indicate the nature of the procedural problem such as missing procedure, procedure inadequate due to technical deficiency, etc.

Personnel errors and human performance related issues may be in the areas of procedures, training, communication, human engineering, management, and supervision. For example, in the area of procedures, errors might be due to missing procedures, procedures which are inadequate due to technical or human factors deficiencies, or which have not been maintained current. In the area of training, errors may be the result of a failure to provide training, having provided inadequate training, or as the result of training (such as simulator training or on-the-job training) that does not provide an environment comparable to that in the plant. Communications errors may be due to inadequate, untimely, misunderstood, or missing communication or due to the quality of the communication equipment. Human engineering issues include those related to the interface or lack thereof between the human and the machine (such as size, shape, location, function or content of displays, controls, equipment or labels) as well as environmental issues such as lighting, temperature, noise, radiation and work area layout. Management errors might be due to management expectations, corrective actions, root cause determinations, or audits which are inadequate, untimely or missing. In the area of supervision, errors may be the result of a lack of supervision, inadequate supervision, job staffing, overtime, scheduling and planning, work practices (such as briefings, logs, work packages, team work, decision making, and housekeeping) or because of inadequate verification, awareness or self-checking.

---

§50.73(b)(2)(ii)(K)

The narrative description must include: "Automatically and manually initiated safety system responses."

The LER should include a discussion of each specific system that actuated or failed to actuate. Do not limit the discussion to ESFs. Indicate the specific equipment that actuated or should have actuated, by train, compatible with EPIX train definitions (e.g., auxiliary feedwater train B). Indicate whether or not the equipment operated successfully.

---

§50.73(b)(2)(ii)(L)

The narrative description must include: "The manufacturer and model number (or other identification) of each component that failed during the event."

The manufacturer and model number (or other identification, such as type, size, or manufacture date) also should be given for each component found failed during the course of the event. An example of other identification could be (for a pipe rupture) size, schedule, or material composition.

(3)   <u>Assessment of Safety Consequences</u>

---

§50.73(b)(3)

The LER shall contain: "An assessment of the safety consequences and implications of the event. This assessment must include the availability of other systems or components that could have performed the same function as the components and systems that failed during the event."

Give a summary assessment of the actual and potential safety consequences and implications of the event, including the basis for submitting the report. Evaluate the event to the extent necessary to fully assess the safety consequences and safety margins associated with the event.

Include an assessment of the event under alternative conditions if the incident would have been more severe (e.g., the plant would have been in a condition not analyzed in its latest SAR) under reasonable and credible alternative conditions, such as a different operating mode. For example, if an event occurred while the plant was at low power and the same event could have occurred at full power, which would have resulted in considerably more serious consequences, this alternative condition should be assessed and the consequences reported.

Reasonable and credible alternative conditions may include normal plant operating conditions, potential accident conditions, or additional component failures, depending on the event. Normal alternative operating conditions and off-normal conditions expected to occur during the life of the plant should be considered. The intent of this section is to obtain the result of the considerations that are typical in the conduct of routine operations, such as event reviews, not to require extraordinary studies.

(4)   <u>Corrective Actions</u>

---

§50.73(b)(4)

The LER shall contain: "A description of any corrective actions planned as a result of the event, including those to reduce the probability of similar events occurring in the future."

Include whether the corrective action was or is planned to be implemented. Discuss repair or replacement actions as well as actions that will reduce the probability of a similar event occurring in the future. For example, "the pump was repaired and a discussion of the event was included in the training lectures." Another example, "although no modification to the instrument was deemed necessary, a caution note was placed in the calibration procedure for the instrument before the step in which the event was initiated."

In addition to a description of any corrective actions planned as a result of the event, describe corrective actions on similar or related components that were done, or are planned, as a direct result of the event. For example, if pump 1 failed during an event and required corrective maintenance and that same maintenance also was done on pump 2, so state.

If a study was conducted, and results are not available within the 30-day period, report the results of the study in a revised LER if they result in substantial changes in the corrective action planned. (See Section 5.1.6 for further discussion of submitting revised LERs.)

(5)   Previous Occurrences

---

§50.73(b)(5)

The LER shall contain: "Reference to any previous similar events at the same plant that are known to the licensee."

The term "previous occurrences" should include previous events or conditions that involved the same underlying concern or reason as this event, such as the same root cause, failure, or sequence of events. For infrequent events such as fires, a rather broad interpretation should be used (e.g., all fires and, certainly, all fires in the same building should be considered previous occurrences). For more frequent events such as ESF actuations, a narrower definition may be used (e.g., only those scrams with the same root cause). The intent of the rule is to identify generic or recurring problems.

The licensee should use engineering judgment to decide how far back in time to go to present a reasonably complete picture of the current problem. The intent is to be able to see a pattern in recurring events, rather than to get a complete 10- or 20-year history of the system. If the event was a high-frequency type of event, 2 years back may be more than sufficient.

Include the LER number(s), if any, of previous similar events. Previous similar events are not necessarily limited to events reported in LERs. If no previous similar events are known, so state. If any earlier events, in retrospect, were significant in relation to the subject event, discuss why prior corrective action did not prevent recurrence.

(6)   LER Text Continuation Sheet (NRC Form 366A)

Use one or more additional text continuation sheets of the LER Form 366A to continue the narrative, if necessary. There is no limit on the number of continuation sheets that may be included.

Drawings, figures, tables, photographs, and other aids may be included with the narrative to help readers understand the event. If possible, provide the aids on the LER form (i.e., NRC Form 366A). In addition, care should be taken to ensure that drawings and photographs are of sufficient quality to permit legible reproduction and micrographic processing. Avoid oversized drawings (i.e., larger than 8 ½ x 11).

5.2.2 Abstract (NRC Form 366, Item 16)

> ## §50.73(b)(1)
>
> The LER shall contain: "A brief abstract describing the major occurrences during the event, including all component or system failures that contributed to the event and significant corrective action taken or planned to prevent recurrence."

Provide a brief abstract describing the major occurrences during the event, including all actual component or system failures that contributed to the event, all relevant operator errors or violations of procedures, the root cause(s) of the major occurrence(s), and the corrective action taken or planned for each root cause. If space does not permit describing failures, at least indicate whether or not failures occurred. Limit the abstract to 1400 characters (including spaces), which is approximately 15 lines of single-spaced typewritten text. Do not use EIIS component function identifiers or the two-letter codes for system names in the abstract.

It is acceptable to describe the entire event in the abstract space. However, the description of the event should be sufficiently detailed so that a knowledgeable reader can understand the complete event. Few reportable events will be so simplistic that they can be adequately described in 1400 characters.

The abstract is generally included in the LER data base to give users a brief description of the event to identify events of interest. Therefore, if space permits, provide the numbers of other LERs that reference similar events in the abstract.

As noted in Section 5.1.10, do not include safeguards, security, or proprietary information in the abstract.

5.2.3 Other Fields on the LER Form

(1) Facility Name (NRC Form 366, Item 1)

Enter the name of the facility (e.g., Indian Point, Unit 1) at which the event occurred. If the event involved more than one unit at a station, enter the name of the nuclear facility with the lowest nuclear unit number (e.g., Three Mile Island, Unit 1).

(2) Docket Number (NRC Form 366, Item 2)

Enter the docket number (in 8-digit format) assigned to the unit. For example, the docket number for Yankee-Rowe is 05000029. Note the use of zeros in this example.

(3) Page Number (NRC Form 366, Item 3)

Enter the total number of pages included (including figures and tables that are attached to Item 17 Text) in the LER package. For continuation sheets, number the pages consecutively beginning with page 2. The LER form, including the abstract and other data is pre-numbered on the form as page 1 of __.

(4)   Title (NRC Form 366, Item 4)

The title should include a concise description of the principal problem or issue associated with the event, the root cause, the result (why the event was required to be reported), and the link between them, if possible. It is often easier to form the title after writing the assessment of the event because the information is clearly at hand.

"Licensee Event Report" should not be used as a title. The title "Reactor Trip" is considered inadequate, because the root cause and the link between the root cause and the result are missing. The title "Personnel Error Causes Reactor Trip" is considered inadequate because of the innumerable ways in which a person could cause a reactor trip. "Technician Inadvertently Injected Signal Resulting in a Reactor Trip" would be a better title.

(5)   Event Date (NRC Form 366, Item 5)

Enter the date on which the event occurred in the eight spaces provided. There are two spaces for the month, two for the day, and four for the year, in that order. Use leading zeros in the first and third spaces when appropriate. For example, June 1, 1987, would be properly entered as 06011987.

If the date on which the event occurred cannot be clearly defined, use the discovery date. See Section 2.11 of this report for further discussion of discovery date.

(6)   Report Number (NRC Form 366, Item 6)

The LER number consists of three parts: (a) the four digits of the event year (based on event date), (b) the sequential report number, and (c) a revision number. The numbering system is shown in the diagram below; the event occurred in the year 1991, it was the 45th event of that year, and the submittal was the 1st revision to the original LER for that event.

Event Year		Sequential Report Number		Revision Number
1991	-	045	-	01

Event Year: Enter the four digits. The event year should be based on the event date (Item 4).

Sequential Report Number: As each reportable event is reported for a unit during the year, it is assigned a sequential number. For example, for the 15th and 33rd events to be reported in a given year at a given unit, enter 015 and 033, respectively, in the spaces provided. Follow the guidelines below to ensure consistency in the sequential numbering of reports.

•   Each unit should have its own set of sequential report numbers. Units at multi-unit sites should not share a set of sequential report numbers.

- The sequential number should begin with 001 for the first event that <u>occurred</u> in each calendar year, using leading zeros for sequential numbers less than 100.

- For an event common to all units of a multi-unit site, assign the sequential number to the lowest numbered nuclear unit.

- If a sequential number was assigned to an event, and it was subsequently determined that the event was not reportable, a "hole" in the series of LER numbers would result. The NRC would prefer that licensees reuse a sequential number rather than leave holes in the sequence. A sequential LER number may be reused even if the event date was later than subsequent reports.

If the licensee chooses not to reuse the number, write a brief letter to the NRC noting that "LER number xxx for docket O5OOOXXX will not be used."

<u>Revision Number</u>: The revision number of the original LER submitted is 00. The revision number for the first revision submitted should be 01. Subsequent revisions should be numbered sequentially (i.e., 02, 03, 04).

(7)   <u>Report Date (NRC Form 366, Item 7)</u>

Enter the date the LER is submitted to the NRC in the eight spaces provided, as described in Section 5.2.4(4) above.

(8)   <u>Other Facilities (NRC Form 366, Item 8)</u>

When a situation is discovered at one unit of a facility that applies to more than the one unit, submit a single LER. LER form items 1, 2, 6, 9, and 10 should refer to the unit primarily affected, or, if both units were affected approximately equally, to the lowest numbered nuclear unit.

The intent of the requirement is to name the facility in which the primary event occurred, whether or not that facility is the lowest numbered of the facilities involved. The automatic use of the lowest number should only apply to cases where both units are affected approximately equally. Item 8 only should indicate the other unit(s) affected. The abstract and the text should describe how the event affected all units.

Enter the facility name and unit number and docket number (see Sections 5.2.4(1) and 5.2.4(2) for format) of any other units at that site that were <u>directly</u> affected by the event (e.g., the event included shared components, the LER described a tornado that threatened both units of a two-unit plant).

(9)   <u>Operating Mode (NRC Form 366, Item 9)</u>

Enter the operating mode of the unit at the time of the event as defined in the plant's technical specifications in the single space provided. For plants that have operating modes such as hot shutdown, cold shutdown, and operating, but do not have numerical operating modes (e.g., Mode 5), place the letter N in Item 9 and describe the operating mode in the text.

(10) Power Level (NRC Form 366, Item 10)

Enter the percent of licensed thermal power at which the reactor was operating when the event occurred. For shutdown conditions, enter 000. For all other operating conditions, enter the correct numerical value (estimate power level if it is not known precisely), using leading zeros as appropriate (e.g., 009 for 9-percent power). Significant deviations in the operating power in the balance of plant should be clarified in the text.

(11) Reporting Requirements (NRC Form 366, Item 11)

Check one or more blocks according to the reporting requirements that apply to the event. A single event can meet more than one reporting criterion. For example: if as a result of sabotage, reportable under §73.71(b), a safety system failed to function, reportable under §50.73(a)(2)(v), and the net result was a release of radioactive material in a restricted area that exceeded the applicable license limit, reportable under §20.2203(a)(3)(i), prepare a single LER and check the three boxes for paragraphs 73.71(b), 50.73(a)(2)(v), and 20.2203(a)(3)(i).

In addition, an event can be reportable as an LER even if it does not meet any of the criteria of 10 CFR 50.73. For example, a case of attempted sabotage (§73.71(b)) that does not result in any consequences that meet the criteria in 50.73 can be reported using the "Other" block. Use the "Other" block if a reporting requirement other than those specified in item 11 was met. Specifically describe this other reporting requirement in the space provided below the "Other" block and in the abstract and text.

(12) Licensee Contact (NRC Form 366, Item 12)

<table>
<tr><td>

§50.73(b)(6)

The LER shall contain: "The name and telephone number of a person within the licensee's organization who is knowledgeable about the event and can provide additional information concerning the event and the plant's characteristics."

</td></tr>
</table>

Enter the name, position title, and work telephone number (including area code) of a person who can provide additional information and clarification for the event described in the LER.

(13) Component Failures (NRC Form 366, Item 13)

Enter the appropriate data for each component failure described in the event.
A failure is defined as the termination of the ability of a component to perform its required function. Unannounced failures are not detected until the next test; announced failures are detected by any number of methods at the instant of occurrence.

If multiple components of the same type failed and all of the information required in Item 13 (i.e., cause, system, component, etc.) was the same for each component, then only a single entry is required in Item 13. Clearly define the number of components that failed in the abstract and text.

The component information elements of this item are discussed below.

Cause: Enter the cause code as shown below. If more than one cause code is applicable, enter the cause code that most closely describes the root cause of the failure.

Cause Code	Classification and Definition
A	Personnel Error is assigned to failures attributed to human errors. Classify errors made because written procedures were not followed or because personnel did not perform in accordance with accepted or approved practice as personnel errors. Do not include errors made as a result of following incorrect written procedures in this classification.
B	Design, Manufacturing, Construction/Installation is assigned to failures reasonably attributed to design, manufacture, construction, or installation of a system, component, or structure. For example, include failures that were traced to defective materials or components otherwise unable to meet the specified functional requirements or performance specifications in this classification.
C	External Cause is assigned to failures attributed to natural phenomena. A typical example would be a failure resulting from a lightning strike, tornado, or flood. Also assign this classification to man-made external causes that originate off site (e.g., an industrial accident at a nearby industrial facility).
D	Defective Procedure is assigned to failures caused by inadequate or incomplete written procedures or instructions.
E	Management/Quality Assurance Deficiency is assigned to failures caused by inadequate management oversight or management systems (e.g., major breakdowns in the licensee's administrative controls, preventive maintenance program, surveillance program, or quality assurance controls, inadequate root cause determination, inadequate corrective action).
X	Other is assigned to failures for which the proximate cause cannot be identified or which cannot be assigned to one of the other classifications.

System: Enter the two-letter system code from Institute of Electrical and Electronics Engineers (IEEE) Standard 805-1984, "IEEE Recommended Practice for System Identification in Nuclear Power Plants and Related Facilities," March 27, 1984. Copies may be obtained from the Institute of Electrical and Electronics Engineers, 345 East 47th Street, New York, NY 10017.

Component: Enter the applicable component code from IEEE Standard 803A-1983, "IEEE Recommended Practice for Unique Identification in Power Plants and Related Facilities - Component Function Identifiers."

Component Manufacturer: Enter the four character alphanumeric reference code. Chapter 18 of the NPRDS Reporting Guidance Manual describes how to access a computerized listing of manufacturer codes maintained on INPO's computer. Designate manufacturers that are not included in the list as X999. If the manufacturer is one used in EPIX, use the manufacturer name as it appears in EPIX.

Reportable to EPIX: Enter a "Y" if the failure is reportable to EPIX and an "N" if it is not reportable.

Include in the LER text and in item 13 of the LER Form any component failure involved in the event, not just components within the scope of EPIX or EIIS.

Failure Continuation Sheet (NRC Form 366B): If more than four failures need to be coded, use one or more of the failure continuation sheets (NRC Form 366B). Code the entries in Items 1, 2, 3, and 6 of the failure continuation sheet to match entries of these items on the initial page of the LER. Complete item 13 in the same manner as item 13 on the basic LER form. Do not repeat failures coded on the basic LER form on the failure continuation sheet. Place any failure continuation sheets after any text continuation sheets and include those sheets in the total number of pages for the LER.

(14) Supplemental Report (NRC Form 366, Item 14)

Check the "Yes" block if the licensee plans to submit a followup report. For example, if a failed component had been returned to the manufacturer for additional testing and the results of the test were not yet available when the LER was submitted, a followup report would be submitted.

(15) Expected Submission Date of Supplemental Report (NRC Form 366, Item 15)

Enter the expected date of submission of the supplemental LER, if applicable. See Section 5.2.4(4) for the proper date format. The expected submission date is a target/planning date; it is not a regulatory commitment.

U.S. NUCLEAR REGULATORY COMMISSION

LICENSEE EVENT REPORT (LER)

(See reverse for required number of
digits/characters for each block)

FACILITY NAME (1)	DOCKET NUMBER (2)	PAGE (3)
	05000	1 OF

TITLE (4)

EVENT DATE (5)			LER NUMBER (6)			REPORT DATE (7)			OTHER FACILITIES INVOLVED (8)	
MONTH	DAY	YEAR	YEAR	SEQUENTIAL NUMBER	REVISION NUMBER	MONTH	DAY	YEAR	FACILITY NAME	DOCKET NUMBER 05000
				--	--				FACILITY NAME	DOCKET NUMBER 05000

OPERATING MODE (9)		THIS REPORT IS SUBMITTED PURSUANT TO THE REQUIREMENTS OF 10 CFR §: (Check one or more) (11)			
		20.2201(b)	20.2203(a)(2)(v)	50.73(a)(2)(i)	50.73(a)(2)(viii)
POWER LEVEL (10)		20.2203(a)(1)	20.2203(a)(3)(i)	50.73(a)(2)(ii)	50.73(a)(2)(x)
		20.2203(a)(2)(i)	20.2203(a)(3)(ii)	50.73(a)(2)(iii)	73.71
		20.2203(a)(2)(ii)	20.2203(a)(4)	50.73(a)(2)(iv)	OTHER
		20.2203(a)(2)(iii)	50.36(c)(1)	50.73(a)(2)(v)	Specify in Abstract below or in NRC Form 366A
		20.2203(a)(2)(iv)	50.36(c)(2)	50.73(a)(2)(vii)	

LICENSEE CONTACT FOR THIS LER (12)

NAME	TELEPHONE NUMBER (Include Area Code)

COMPLETE ONE LINE FOR EACH COMPONENT FAILURE DESCRIBED IN THIS REPORT (13)

CAUSE	SYSTEM	COMPONENT	MANUFACTURER	REPORTABLE TO EPIX		CAUSE	SYSTEM	COMPONENT	MANUFACTURER	REPORTABLE TO EPIX

SUPPLEMENTAL REPORT EXPECTED (14)			EXPECTED SUBMISSION DATE (15)	MONTH	DAY	YEAR
YES (If yes, complete EXPECTED SUBMISSION DATE).		NO				

ABSTRACT (Limit to 1400 spaces, i.e., approximately 15 single-spaced typewritten lines) (16)

REQUIRED NUMBER OF DIGITS/CHARACTERS
FOR EACH BLOCK

BLOCK NUMBER	NUMBER OF DIGITS/CHARACTERS	TITLE
1	UP TO 46	FACILITY NAME
2	8 TOTAL 3 IN ADDITION TO 05000	DOCKET NUMBER
3	VARIES	PAGE NUMBER
4	UP TO 76	TITLE
5	8 TOTAL 2 FOR MONTH 2 FOR DAY 4 FOR YEAR	EVENT DATE
6	9 TOTAL 4 FOR YEAR 3 FOR SEQUENTIAL NUMBER 2 FOR REVISION NUMBER	LER NUMBER
7	8 TOTAL 2 FOR MONTH 2 FOR DAY 4 FOR YEAR	REPORT DATE
8	UP TO 18 -- FACILITY NAME 8 TOTAL -- DOCKET NUMBER 3 IN ADDITION TO 05000	OTHER FACILITIES INVOLVED
9	1	OPERATING MODE
10	3	POWER LEVEL
11	1 CHECK BOX THAT APPLIES	REQUIREMENTS OF 10 CFR
12	UP TO 50 FOR NAME 14 FOR TELEPHONE	LICENSEE CONTACT
13	CAUSE VARIES 2 FOR SYSTEM 4 FOR COMPONENT 4 FOR MANUFACTURER EPIX VARIES	EACH COMPONENT FAILURE
14	1 CHECK BOX THAT APPLIES	SUPPLEMENTAL REPORT EXPECTED
15	8 TOTAL 2 FOR MONTH 2 FOR DAY 4 FOR YEAR	EXPECTED SUBMISSION DATE

NUREG-1022, Rev. 1

FACILITY NAME (1)	DOCKET	LER NUMBER (6)			PAGE (3)
		YEAR	SEQUENTIAL NUMBER	REVISION NUMBER	
	05000		--	--	OF

TEXT *(If more space is required, use additional copies of NRC Form 366A)* (17)

NRC FORM 366A (mm-yyyy)

FACILITY NAME (1)	DOCKET	LER NUMBER (6)			PAGE (3)
	05000	YEAR	SEQUENTIAL NUMBER	REVISION NUMBER	OF
			--	--	

COMPLETE ONE LINE FOR EACH COMPONENT FAILURE DESCRIBED IN THIS REPORT (13)

CAUSE	SYSTEM	COMPONENT	MANUFACTURER	REPORTABLE TO EPIX		CAUSE	SYSTEM	COMPONENT	MANUFACTURER	REPORTABLE TO EPIX

APPENDIX A

HISTORICAL PERSPECTIVE ON EVENT REPORTING

Origin of 10 CFR 50.72 and 50.73

In December 1980, the U.S. Nuclear Regulatory Commission (NRC) determined that requirements for reporting operational experience data needed major revision and approved the development of an integrated operational experience reporting (IOER) system. The IOER system was to combine, modify, and make mandatory the existing licensee event report (LER) system and the industry supported, voluntary nuclear plant reliability data system (NPRDS). The NPRDS contains both engineering and failure data submitted by nuclear power plant licensees on specified plant components and systems. An advance notice of proposed rulemaking concerning the IOER system was published on January 15, 1981 (46 FR 3541).

On June 8, 1981, the Institute of Nuclear Power Operations (INPO) stated it would assume responsibility for managing and funding the NPRDS and would audit member utilities to assess the adequacy of their participation in the NPRDS. The NRC believed the NPRDS would provide the necessary operating experience data and further development of the IOER system was discontinued.

On May 6, 1982, the NRC published a notice of proposed rulemaking in the *Federal Register* (47 FR 19543) that would modify and codify the existing LER system. On July 26, 1983, after consideration of public comments, the NRC published in the *Federal Register* (48 FR 33850) a final rule under 10 CFR 50.73, which modified and codified the LER system and became effective on January 1, 1984. In the rule, the Commission clearly indicated that the NPRDS is a vital adjunct to 10 CFR 50.73 for component data.[1] The purpose of the rule was to standardize the reporting requirements for all nuclear power plant licensees, to eliminate reporting events of low individual significance, and to require more thorough documentation and analyses of reported events. Licensees are to submit such reports within 30 days of discovery. The revised system also permits licensees to use the LER procedures for various other reports required under specific sections of 10 CFR Part 20 and Part 50.

Also effective January 1, 1984, the NRC amended its immediate notification requirements of significant events at operating nuclear power reactors (10 CFR 50.72) to clarify reporting criteria and to require early reports only on those matters of value to the exercise of the Commission's responsibilities. The amended rule was published in the *Federal Register* (48 FR 39039) on August 29, 1983, and corrections to the rule (48 FR 40882) were published on September 12, 1983. Among the changes made were the use of terminology, phrasing, and reporting thresholds similar to those of 10 CFR 50.73 whenever possible. Therefore, most events reported under 10 CFR 50.72 also will require an in-depth followup report under 10 CFR 50.73.

NRC Workshops and Event Reporting Guidelines

In September 1983, the NRC staff published NUREG-1022, "Licensee Event Reporting System," to provide supporting information and guidelines to persons responsible for the

---

[1] On January 1, 1997 NPRDS was replaced by a new reporting system entitled Equipment Performance and Information Exchange (EPIX).

preparation and review of LERs. NUREG-1022 includes (1) a brief description of how the NRC analyzes LERs, (2) a restatement of the guidance contained in the Statements of Consideration that accompanied the publication of the LER rule, (3) a set of examples of potentially reportable events with staff comments on the actual reportability of each event, (4) guidelines on how to prepare an LER and use the LER form, and (5) guidelines on submittal of LERs.

Between October 25 and November 16, 1983, the NRC held five regional workshops to discuss the new LER rule (10 CFR 50.73) and the revised emergency notification rule (10 CFR 50.72). Supplement 1 to NUREG-1022 was published in February 1984 to provide a summary of answers to questions asked during the workshops.

Supplement 2 to NUREG-1022, issued in September 1985, contained evaluations of the quality and completeness of an industry-wide sample of 415 LERs. The study was performed for the NRC Office for Analysis and Evaluation of Operational Data (AEOD) by EG&G, Inc., at Idaho National Engineering Laboratory. The report identifies deficiencies in LER content and recommends corrective actions.

NRC Regulatory Impact Study (Draft NUREG-1395)

In the fall of 1989, the NRC staff surveyed personnel from 13 nuclear power utilities to obtain their views on the potential effect that NRC regulatory activities were having on the safe operation of their nuclear plants. This survey was documented in NUREG-1395, "Industry Perceptions of the Impact of the U.S. Nuclear Regulatory Commission on Nuclear Power Plant Activities," Draft, March 1990. Section 8, "Reporting Events," of NUREG-1395 included industry comments on reporting required by 10 CFR 50.72 and 50.73.

Specific industry concerns included the need for reporting

- inadvertent actuations of engineered safety feature (ESF) equipment

- actuation of ESF equipment involving no safety significance

- plant shutdowns required by plant technical specifications even though the action statements of the technical specifications were being met

- grass fires not affecting plant safety

- radiation exposures in excess of regulatory limits

Amendment of 10 CFR 50.72 and 50.73

On September 10, 1992 the NRC published a final rule in the *Federal Register* (57 FR 41373) to eliminate the requirement to report certain events which had been determined to be of little or no safety significance. These were events that resulted in invalid actuation of several specific engineered safety features.

Other amendments have been made from time to time in order to make these sections conform with changes in other sections. For example, the requirements for reporting releases of radioactive material have been amended to conform with changes to 10 CFR Part 20.

Revision of NUREG-1022

Partially in response to the industry's concerns regarding event reporting described in NUREG-1395, the NRC sponsored four additional regional workshops on event reporting during September to November 1990.

The NRC staff determined that additional clarification was needed to further improve the usefulness, quality, and threshold of reporting by the licensees under 10 CFR 50.72 and 50.73. Therefore, a draft of Revision 1, to NUREG-1022, to encompass and supersede NUREG-1022 and Supplements 1 and 2, was published for comment in September 1991. In May 1992 and again in May 1993, public meetings were held to discuss the issues raised by public comments. After addressing these issues, a second draft was published for comment in February 1994.

The intent of this Revision 1 is to clarify reporting requirements of 10 CFR 50.72 and 50.73 without changing those reporting requirements. Accordingly, most of the guidance is not new or different from generic reporting guidance previously published in the statements of considerations for the rules, NUREG-1022, its Supplements 1 and 2, or generic correspondence such as generic letters and information notices. This final version of Revision 1 has been modified as appropriate to address public comments on the second draft that was published for comment in February 1994. Further, at the staff's initiative the instructions for preparation of LERs were augmented to address consistency of information provided in LERs which is used to understand events, as discussed in Sections 2.8 and 5.

APPENDIX B

EMERGENCY NOTIFICATION SYSTEM PROCESS

NRC Prompt Response Personnel

Headquarters Operations Officer

The U.S. Nuclear Regulatory Commission (NRC) Operations Center is continuously staffed with an NRC headquarters operations officer (HOO), who holds a degree in engineering and works for the Office for Analysis and Evaluation of Operational Data (AEOD). HOOs are trained to receive licensee notifications via the emergency notification system (ENS) made under Title 10 of the *Code of Federal Regulations* (10 CFR) Section 50.72. In addition, they are trained to receive materials, security or transportation events, as well as inquiries from the public or media. A second HOO is usually on duty during normal working hours to help with the more frequent communications experienced during the work day.

Each HOO has previous nuclear experience and receives extensive classroom and simulator training on both boiling-water and pressurized-water reactor systems at the NRC Technical Training Center.

Although HOOs have a good general understanding of nuclear power plants, they do not have expert knowledge of each specific plant. The HOOs ask questions and rely on the licensees to explain plant-specific details, terms, and the limiting conditions for operation of related technical specifications, to ensure they understand the significance of the event and are able to answer pertinent questions. The HOOs will attempt to obtain all of the details of the event that bear on its safety significance, even if those details would not otherwise be reportable.

The HOO determines, by procedure, how quickly the ENS event information needs to be disseminated to various NRC officials and other Federal agencies and prepares a written report of the oral ENS notification (ENS Event Notification Report) for electronic distribution to the NRC Office of Nuclear Reactor Regulation (NRR), NRC regional offices and the Institute of Nuclear Power Operations, by 7:30 a.m. each weekday morning.

Emergency Officer

If an emergency is declared or if it appears that the event may have significant plant-specific or generic interest to the NRC, the HOO notifies the emergency officer (EO). The EO is assigned on a weekly rotation from NRC staff members of the Senior Executive Service, and is on call 24 hours per day. These are typically NRR division directors, assistant division directors, or branch chiefs, who are responsible for the NRC response to an event. The EO decides which other NRC managers should be informed to participate in responding to the event. The EO also participates in deciding whether the NRC Operations Center and/or the applicable NRC regional incident response center will be partially or fully staffed to continuously monitor the event.

Regional Duty Officer

The HOO promptly informs the regional duty officer (RDO) of any ENS notification affecting the RDO's NRC region. The RDO, who is a senior NRC employee (typically a branch chief or division director) in the applicable NRC region, is assigned a weekly rotation and is on call 24 hours per day. The RDO informs the responsible NRC section chief and other NRC staff, as needed. The NRC regional staff follow up on the plant-specific aspects of each event through the responsible section chief, resident inspectors, and other NRC managers or technical experts, as needed.

Resident Inspector

If the safety significance of an event warrants or if the HOO can not obtain a clear understanding of an event, the RDO may request a resident inspector to immediately investigate, monitor, and report back to the NRC region and headquarters on the situation.

Licensees are encouraged to work with a resident inspector if they have a question regarding the reportability of an issue. If the resident inspector cannot provide guidance, he or she can direct the licensee through the region to headquarters for a more definitive discussion. The resident inspector will not make the decision, but can advise what the regulations require. The resident inspector should be informed about an event whenever an ENS notification is made.

The NRC relies on the continuously staffed NRC Operations Center, not the resident inspector, to notify the appropriate NRC staff of a reportable event.

NRC Response to ENS Notifications

NRC Response Options

There is a wide range of typical NRC headquarters and region responses to an ENS notification, depending on the safety significance of the event, including:

- The NRC Operations Center and the NRC regional incident response center may be fully activated and a site team sent to the plant.

- Specific NRC staff may monitor the progress of the event from the NRC Operations Center and/or regional incident response center and an NRC team may be sent to the plant.

- A resident inspector may be requested to immediately investigate, monitor, and report back to the NRC region and/or headquarters.

- Conference calls among NRC headquarters, region, and licensee management may be established.

- The EO, RDO, and HOO may follow the progress of the event and request specific information from the licensee on a periodic basis until the plant is in a safe condition.

- The RDO may receive the notification and contact the resident inspector for additional information.

Additional NRC Operating Event Review

Each working day the NRR Events Assessment and Generic Communications Branch (PECB) and the AEOD Reactor Analysis Branch (RAB) obtain copies of notifications of events that were received in the NRC Operations Center since the beginning of the previous working day. Copies of the daily report from each regional office also are obtained. These reports present the results of the regional offices' review of events occurring within the region since the previous working day, regardless of whether licensees have submitted notifications under 10 CFR 50.72.

Each working day PECB and RAB personnel screen the notifications and regional daily reports to identify events that are potentially significant. A telephone conference follows at a preset time in the morning among representatives of PECB, RAB, the NRC Operations Center, and others. The conference call is made to discuss the significance of the events and identify specific events for further assessment. If an assessment is needed, engineers are assigned to determine what happened during the event, what caused the event, what the consequences might be, what corrective or preventive action is being taken, and whether that action is sufficient. If the event is still ongoing, then the engineer follows its development.

During assessment of the event, the assigned engineer determines whether the event is generic, significant, or both. The event is generic if other nuclear power plants have the potential for occurrence of a similar event. Searches of plant operational experience data bases may be performed by RAB personnel to identify similar occurrences and assess generic applicability. The event is significant if any of the following occurred:

- potential or actual degradation occurred in safety-related equipment or structures, fuel integrity, the primary coolant pressure boundary, or containment
- release of radioactivity (in excess of 10 CFR Part 20 limits) occurred
- the plant was operated outside technical specification limits
- a scram with complications occurred
- other conditions warranted attention by NRC

If the event is classified as significant, senior NRC management are informed at the next weekly events briefing meeting. Briefing information, including event summaries and diagrams, are placed in the Public Document Room (PDR). The event also is entered into the PECB significant event tracking system. Each quarter the significant events are compiled and published in the NRC performance indicator report ("Performance Indicators for Operating Commercial Nuclear Power Reactors," issued by AEOD and available in the NRC PDR).

Additional event followup actions performed by NRR, the appropriate NRC regional office, and AEOD personnel may include consulting with the Executive Director for Operations in the selection of an incident investigation team (IIT), participating in the decision to dispatch an augmented inspection team (AIT) to the site and in the selection of the team members, or performing a human performance evaluation at the plant. The appropriate NRC regional office has the direct responsibility for routine followup and inspection related to reportable events.

Depending on the number or types of event notifications by licensees, NRR also may issue NRC generic letters, bulletins, and information notices.

APPENDIX C

LICENSEE EVENT REPORT REVIEW PROGRAMS

Title 10 of the *Code of Federal Regulations* (10 CFR) Section 50.73 specifies that licensee event reports (LERs) shall include a detailed narrative description of reportable operating experience, including safety significant and potentially safety significant events and conditions. By describing in detail the events or conditions required to be reported, LERs provide information for detailed studies of events or conditions that might affect the health and safety of the public.

Variations in LER counts from plant to plant can result from numerous factors, only one of which is an actual difference in safety performance. Thus, the number of LERs submitted by a plant should not be used as a measure of the plant's safety performance.

In addition to prompt followup to ENS notifications described in Appendix B, longer-term followup of licensee events is conducted using the LER information. The appropriate U.S. Nuclear Regulatory Commission (NRC) regional office conducts plant-specific followup, the Office of Nuclear Reactor Regulation (NRR) conducts plant-specific and generic reviews, and the Office for Analysis and Evaluation of Operational Data (AEOD) and its contracted national laboratories, screen, classify, categorize, trend, assess, and store the data for each LER. Those events and conditions, both plant-specific and generic, that appear to be important to safety are further analyzed or evaluated. From this review process, the NRC determines further actions such as (1) a special study initiated to propose revisions to regulatory programs, (2) reporting as an abnormal occurrence to Congress, or (3) dissemination to the U. S. nuclear power industry through generic communications and to the international community through the Nuclear Energy Agency (NEA). The NEA is part of the Organization for Economic Cooperation and Development and gathers information from its member countries on the operating experience of commercial nuclear power plants worldwide.

Several fundamental objectives associated with the LER analysis process are

- to identify and quantify events and conditions that are precursors to potential severe core damage

- to discover emerging trends or patterns of potential safety significance

- to identify events that are important to safety and their associated safety concerns and root causes and to determine the adequacy of corrective actions taken to address the safety concerns

- to assess the generic applicability of events

A precursor to potential severe core damage is an event or condition that could have been serious if plant conditions, personnel action, or the extent of equipment failure or faulting had been slightly different than that which occurred.

An analysis of trends and patterns in operational experience identifies repetitive events and failures and searches past operating history for similar events and failures to determine if the frequency of such events or failures is significant enough to be a cause for concern. When appropriate, an NRC bulletin or information notice is issued or a generic study initiated to focus on the nature, cause, consequences, and possible corrective actions of such a situation. Trends and patterns analysis usually applies to events and conditions that individually are of low safety significance but that become a safety significant factor because of repetition or, more accurately, the frequency of occurrence.

AEOD studies of events that are important to safety are documented in the following reports:

- Case study reports document substantive, in-depth analyses of safety issues and the bases for AEOD recommendations for regulatory or industry actions.

- Special study reports document accelerated assessments of significant operating events and contain recommendations for remedial actions, if appropriate.

- Engineering evaluation reports document assessments of significant operating events and contain suggestions for remedial actions, if appropriate.

- Technical review reports document studies of issues that were determined to have little safety significance.

AEOD uses the sequence coding and search system (SCSS) data base for storage and retrieval of LER data. This system, developed in the early 1980's and maintained under contract at the Oak Ridge National Laboratory, Oak Ridge, Tennessee, contains an average of 150 items of information in its data base for each LER submitted since 1980.

AEOD uses LER data from the SCSS data base to support NRC activities such as plant diagnostic evaluations, NRC senior management meetings, and performance indicators. The SCSS data base also is a primary source of information for AEOD studies. In addition, NRC's Office of Nuclear Regulation, Office of Nuclear Regulatory Research, and regional offices use the SCSS as a source of information on operating experience.

AEOD also maintains LER information in the trends and patterns data base at Idaho National Engineering and Environmental Laboratory (INEEL). This data base supports such specific AEOD studies as those covering performance indicator data for reactor trips, safety system actuations, and safety system failures. The INEL data base also is used to calculate forced outage rates and equipment-forced outages per 1000 critical hours, as well as to support the preparation of Commission site visit briefing packages, special studies, and the evaluations of selected plants.

The information from LERs is widely used within the nuclear industry, both nationally and internationally. For example, the industry's Institute of Nuclear Power Operation (INPO) uses LERs as a basis for providing operational safety experience feedback data to individual utilities through such documents as significant operating experience reports, significant event reports, significant event notifications, and operations and maintenance reminders. U.S. vendors and nuclear steam system suppliers, as well as other countries and international organizations, use LER data as a source of operational experience data.

APPENDIX D

10 CFR 50.72 INCLUDING
STATEMENT OF CONSIDERATIONS

Published in the *Federal Register*
On August 29, 1983
(Vol. 48, No. 168, pages 39039-39046)

NOTE: This *Federal Register* notice does not provide a current version of 10 CFR 50.72, which has been amended several times since 1983. Its purpose here is to present the Statement of Considerations, which explains the basic reporting requirements of 10 CFR 50.72.

10 CFR Part 50

Immediate Notification Requirements of Significant Events At Operating Nuclear Power Reactors

AGENCY: Nuclear Regulatory Commission.

ACTION: Final rule.

SUMMARY: The Nuclear Regulatory Commission is amending its regulations which require timely and accurate information from licensees following significant events at commercial nuclear power plants. Experience with existing requirements and public comments on a proposed revision of the rule indicate that the existing regulation should be amended to clarify reporting criteria and to require early reports only on those matters of value to the exercise of the Commission's responsibilities. The amended regulation will clarify the list of reportable events and provide the Commission with more useful reports regarding the safety of operating nuclear power plants.

EFFECTIVE DATE: January 1, 1984.

FOR FURTHER INFORMATION CONTACT: Eric W. Weiss, Office of Inspection and Enforcement, U.S. Nuclear Regulatory Commission, Washington, D.C. 20555; Telephone (301) 492–4973.

SUPPLEMENTARY INFORMATION:

I. Background

On February 29, 1980, the Commission amended its regulations without prior notice and comment to require timely and accurate licensee reporting of information following significant events at operating nuclear power reactors (45 FR 13434). The purpose of the rule was to provide the Commission with immediate reporting of twelve types of significant events where immediate Commission action to protect the public health and safety may be required or where the Commission needs accurate and timely information to respond to heightened public concern. Although the rule was made immediately effective, comments were solicited. Many commenters believed the rule was in some respects either vague and ambiguous or overly broad.

After obtaining experience with notifications required by the rule, the Commission published in the Federal Register a notice of proposed rulemaking on December 21, 1981 (46 FR 61894) and invited public comment. The proposal was made to meet two objectives: change 10 CFR 50.54 to implement Section 201 of the NRC's 1980 Fiscal Year Authorization Act and change 10 CFR 50.72 to more clearly specify the significant events requiring licensees to immediately notify NRC.

The problems and issues which this rulemaking addresses and the solutions that it provides can be summarized in five broad areas:

1. Authorization Act for FY80

Section 201 of the Nuclear Regulatory Commission Authorization Act for Fiscal Year 1980 (Pub. L. 96–295) provides:

(a) Section 103 of the Atomic Energy Act of 1954 is amended by adding at the end thereof the following new subsection: f. Each license issued for a utilization facility under this section or section 104b. shall require as a condition thereof that in case of any accident which could result in an unplanned release of quantities of fission products in excess of allowable limits for normal operation established by the Commission, the licensee shall immediately so notify the Commission. Violation of the condition prescribed by this subsection may, in the Commission's discretion, constitute grounds for license revocation. In accordance with section 187 of this Act, the Commission shall promptly amend each license for a utilization facility issued under this section or section 104b. which is in effect on the date of enactment of this subsection to include the provisions required under this subsection.

Accordingly, this rulemaking includes an amendment to 10 CFR 50.54 that would add an appropriate notification requirement as a condition in the operating license of each nuclear utilization facility licensed under section 103 or 104b. of the Atomic Energy Act of 1954, as amended, 42 U.S.C. 2133, 2134b. These facilities generally are the commercial nuclear power facilities which produce electricity for public consumption. Research and test reactors are not subject to the license condition as they are licensed under section 104a. or 104c. of the Act. Under the amendment to 10 CFR 50.54, licensees falling under sections 103 or 104b. would be required, as a condition of their respective operating licenses, to notify the NRC immediately of events specified in 10 CFR 50.72.

2. Unnecessary Reports

Several categories of reports required by § 50.72 are not useful to the NRC. Among these categories are reports of: worker injury, small radioactive releases, and minor security problems. For example, reports are presently required if a worker onsite experience chest pains or another illness not related to radiation and is sent to a hospital for evaluation; or if the vent stack monitor moves upward a few percent yet radiation levels remain 100,000 times below technical specification limits; or if the security computer malfunctions for a few minutes.

This rulemaking eliminates such reporting requirements from § 50.72 and in general clarifies and narrows the scope of reporting. However, revision of Part 73 of the Commission's regulations is necessary to resolve all problems with security reports.

3. Terminology, Phrasing, and Reporting Thresholds

The various sections of 10 CFR 50 have different phrasing, terminology, and thresholds in the reporting criteria. Even when no different meaning is intended a change in wording can cause confusion.

This rulemaking has been carefully written to use terminology, phrasing, and reporting thresholds that are either identical to or similar to those in § 50.73, whenever possible. Other conforming amendments to Parts 20, 21, 73, and in § 50.55 and Appendix E of Part 50 are under development.

As a parallel activity to the preparation of § 50.72, on July 26, 1983, the Commission has published a Licensee Event Report (LER) Rule (§ 50.73) which requires licensees for operating nuclear power plants to prepare detailed written reports for certain events (48 FR 33850).

4. Coordination with Licensee's Emergency Plan

The current scheme for licensees' emergency plans includes four Emergency Classes. When the licensee declares one of the four Emergency Classes, it must report this to the Commission as required by § 50.72. The lowest of the four Emergency Classes, Notification of Unusual Event, has resulted in unnecessary emergency declarations. Events that fall within the Unusual Event class have been neither emergencies in themselves nor precursors of more serious events that are emergencies.

Although changes to the definition of the Emergency Classes are not being made in this rulemaking, a new reporting scheme that would ultimately eliminate "Unusual Event" as an Emergency Class requiring notification can be adopted consistent with this rule. A proposed rulemaking which would redefine the Emergency Classes in § 50.47 is in preparation and may soon be published for public comment. This final rulemaking makes possible the elimination of "Unusual Event" as an emergency class without further amendment of § 50.72 by including in the category of non-Emergencies the subcategory of "one-hour reports."

5. Vague or Ambiguous Reporting Criteria

The reporting criteria in § 50.72 have been revised in order to clarify their scope and intent. The criteria were revised for the proposed rule and in response to public comment. The "Analysis of Comments" portion of this Federal Register notice describes in more detail specific examples of changes in wording intended to eliminate vagueness or ambiguity.

II. Analysis of Comments

Twenty letters of comment were received in response to the Federal Register notice published on December 21, 1981 (46 FR 61894).[1] Of the twenty letters of comment received, the vast majority (15 of 20) were from utilities owning or operating nuclear power plants. This Federal Register notice described the proposed revision of 10 CFR 50.72, "Notification of Significant Events," and 10 CFR 50.54, "Conditions of Licenses." A discussion of the more significant comments follows:

---

[1] Copies of these documents are available for public inspection and copying for a fee in the NRC Public Document Room, 1717 H Street, N.W., Washington, D.C. 20555

Conditions of Licenses (§ 50.54)

A few commenters said that the "Commission already has the ability to enforce its regulations and does not need to incorporate the items as now proposed into conditions of license."

The Commission has decided to promulgate the proposed revision of § 50.54, "Conditions of Licenses," in order to satisfy the intent of Congress as expressed in Section 201 of the Nuclear Regulatory Commission Authorization Act for Fiscal Year 1980. This Act and its relationship to § 50.54 are discussed in detail in the Federal Register notice for the proposed rule (46 FR 61894).

Coordination With Other Reporting Requirements (Final Rule § 50.72)

Seven commenters said that the NRC should coordinate the requirements of 10 CFR 50.72 with other rules, with NUREG–0654, "Criteria for Preparation and Evaluation of Radiological Emergency Response Plans and Preparedness in Support of Nuclear Plants," and with Regulatory Guide 1.16, "Reporting of Operating Information . . ." Many of these letters identified overlap, duplication, and inconsistency among NRC's reporting requirements.

The Commission is making a concerted effort to ensure consistent and coordinated reporting requirements. The requirements contained in the revision of 10 CFR 50.72 are being coordinated with revision of § 50.73, § 50.55(e), Appendix E of Part 50, § 20.402, § 73.71, and Part 21.

Citing 10 CFR 50.72 as a Basis for Notification (Final Rule § 50.72(a)(4))

A few commenters objected to citing § 50.72 as a basis when making a telephone notification. The letters of comment questioned the purpose, legal effect, and burden on the licensee.

The Commission does not believe that it is an unnecessary burden for a licensee to know and identify the basis for a telephone notification required by § 50.72. There have been many occasions when a licensee could not tell the NRC whether the telephone notification was being made in accordance with Technical Specifications, 10 CFR 50.72, some other requirement, or was just a courtesy call. Unless the licensee can identify the nature of the report, it is difficult for the NRC to know what significance the licensee attaches to the report, and it becomes more difficult for the NRC to respond quickly and properly to the event.

Immediate Shutdown (Final Rule § 50.72(b)(1)(i))

Several commenters objected to the use of the term, "immediate shutdown," saying that Technical Specifications do not use such a term.

The term is used in some but not all Technical Specifications. Consequently, the Commission has revised the reporting criterion in question. The final rule requires a report upon the initiation of any nuclear power plant shutdown required by Technical Specifications.

Plant Operating and Emergency Procedures (Final Rule § 50.72(b)(1)(ii))

Several commenters said that the reporting criteria should not make reference to plant operating and emergency procedures because:

a. It would take operators too long to decide whether a plant condition was covered by the procedures,

b. The procedures cover events that are not of concern to the NRC, and

c. The procedures vary from plant to plant.

While the plant operating personnel should be familiar with plant procedures, it is true that procedures vary from plant to plant and cover events other than those which compromise plant safety. However, the wording of the reporting criteria has been modified (§ 50.72(b)(1)(ii) in the final rule) to narrow the reportable events to those that significantly compromise plant safety. Notwithstanding the fact that the procedures vary from plant to plant, the Commission has found that this criterion results in notifications indicative of serious events. The narrower, more specific wording will make it possible for plant operating personnel to identify reportable events under their specific operating procedures.

Building Evacuation (Final Rule § 50.72(b)(1)(iii))

Ten commenters said that the proposed § 50.72(b)(6)(iii) regarding "any accidental, unplanned or uncontrolled release resulting in evacuation of a building" was unclear and counterproductive in that it could cause reluctance to evacuate a building. Many of these commenters stated that the reporting of in-plant releases of radioactivity that require evacuation of individual rooms was inconsistent with the general thrust of the rule to require reporting of significant events. They noted that minor spills, small gaseous waste releases, or the disturbance of contaminated particulate matter (e.g., dust) may all require the temporary evacuation of individual rooms until the

airborne concentrations decrease or until respiratory protection devices are utilized. They noted that these events are fairly common and should not be reportable unless the required evacuation affects the entire facility or a major part of it.

The Commission agrees. The wording of this criterion has been changed to include only those events which significantly hamper the ability of site personnel in performance of duties necessary for safe operation.

One commenter was concerned that events occurring on land owned by the utility adjacent to its plant might be reportable. This is not the intent of this reporting requirement. The NRC is concerned with the safety of plant and personnel on the utility's site and not with non-nuclear activities on land adjacent to the plant.

Explicit Threats (Final Rule § 50.72(b)(1)(vi))

A few commenters said that the intent of the term. "explicitly threatens," was unclear. Those commenting wondered what level of threat was involved. The term. "explicitly threatens, " has been deleted from the final rule. Instead, the final rule refers to "any event that poses an actual threat to the safety of the nuclear power plant" [§ 50.72(b)(1)(vi)] and gives examples so that it is clear the Commission is interested in real or actual threats as opposed to threats without credibility.

Notification Timing (Final Rule § 50.72(b)(2))

The commenters generally had two points to make regarding the timing of reports to the NRC. First, the comments supported notification of the NRC after appropriate State or local agencies have been notified. Second, two commenters requested a new four-to six-hour report category for events not warranting a report with one hour.

Allowing more time for reporting some non-Emergency events would lessen the impact of reporting on the individuals responsible for maintaining the plant in a safe condition. Limiting the extension of the deadline to four hours ensures that the report is made when the information is fresh in the minds of those involved and that it is more likely to be made by those involved rather than by others on a later shift.

Other, more significant non-Emergency events and all declarations of an Emergency must continue to be reported within one hour. The one-hour deadline is necessary if the Commission is to fulfill its responsibilities during and following the most serious events

occurring at operating nuclear power plants. A deadline shorter than one hour was not adopted because the Commission does not want to interfere with the operator's ability to deal with an accident or transient in the first few critical minutes.

Therefore, based on these comments and its experience the NRC has established a "four-hour report," as was suggested.

Reactor Scrams (Final Rule § 50.72(b)(2)(ii))

Several commenters said that reactor scrams, particularly those scrams below power operation, should not require notification of the NRC within one hour.

In response to these comments, the Commission had changed the reporting deadline to four hours. However, the Commission does not regard reactor scrams as "non-events," as stated in some letters of comment. Information related to reactor scrams has been useful in identifying safety-related problems. The Commission agrees that four hours is an appropriate deadline for this reporting requirement because these events are not as important to immediate safety as are some other events.

Radioactive Release Threshold (Final Rule § 50.72(b)(2)(iv))

Several commenters said that the threshold of 25% of allowable limits for radioactive releases was too low for one-hour reporting.

Based upon these comments and its experience, the Commission has changed the threshold of reporting to those releases exceeding two times Part 20 concentrations when averaged over a period of one hour. This will eliminate reports of releases that represent negligible risk to the public.

The Commission has found that low level radioactive releases below two times Part 20 concentrations do not, in themselves, warrant immediate radiological response.

This paragraph requires the reporting of those events that cause an unplanned or uncontrolled release of a significant amount of radioactive material to offsite areas. Unplanned releases should occur infrequently; however, when they occur, at least moderate defects have occurred in the safety design or operational control established to avoid their occurrence and, therefore, these events should be reported.

Personnel Radioactive Contamination (Final Rule § 50.72(b)(2)(v))

Several commenters objected to the use of vague terms such as "extensive onsite contamination" and "readily removed" in one of the reporting criteria of the proposed rule.

Based on this comment, new criteria have been prepared that use more specific terms. For example, one new criterion requires reporting of "Any event requiring the transport of a radioactively contaminated person to an offsite medical facility for treatment." Experience with telephone notifications made to the NRC Operations Center suggests that this new criterion will be easily understood.

III. Paragraph-by-Paragraph Explanation of the Rule

Paragraph 50.72(a) reflects some consolidation of language that was repeated in various subparagraphs of the proposed rule. In general, the intent and scope of this paragraph do not reflect any change from the proposed rule.

Several titles were added to this and subsequent sections. For example, paragraph 50.72(b) is titled "Non-Emergency Events" and it has two subparagraphs: (b)(1), titled, "One-Hour Reports" and (b)(2), "Four-Hour Reports." The events which have a one-hour deadline are those having the potential to escalate to an Emergency Class. The four-hour deadline is explained in the analysis of paragraph (b)(2).

Paragraph 50.72(b)(1)(i)(A) requires reporting of "The initiation of any nuclear plant shutdown required by Technical Specifications." Although the intent and scope have not changed, the change in wording between the proposed and final rule is intended to clarify that prompt notification is required once a shutdown is initiated.

In response to public comment, the term "immediate shutdown" that was used in the proposed rule is not used in the final rule. The term was vague and unfamiliar to those licensees who did not have Technical Specifications using the term.

This reporting requirement is intended to capture those events for which Technical Specifications require the initiation of reactor shutdown. This will provide the NRC with early warning of safety significant conditions serious enough to warrant shutdown of the plant.

Paragraph 50.72(b)(1)(i)(B) was added to be consistent with existing requirements in § 50.54(x) and the existing § 50.72(c) as published in the Federal Register on April 1, 1983 (48 FR 13966) which require the licensee to notify the NRC Operations Center by telephone when the licensee departs

from a license condition or technical specification.

Paragraph 50.72(b)(1)(ii), encompassing events previously classified as Unusual Events and some events captured by proposed § 50.72(b)(1) was added to provide for consistent, coordinated reporting requirements between this rule and 10 CFR 50.73 which has a similar provision. Public comment suggested that there should be similarity of terminology, phrasing, and reporting thresholds between § 50.72 and § 50.73. The intent of this paragraph is to capture those events where the plant, including its principal safety barriers, was seriously degraded or in an unanalyzed condition. For example, small voids in systems designed to remove heat from the reactor core which have been previously shown through analysis not to be safety significant need not be reported. However, the accumulation of voids that could inhibit the ability to adequately remove heat from the reactor core, particularly under natural circulation conditions, would constitute an unanalyzed condition and would be reportable. In addition, voiding in instrument lines that results in an erroneous indication causing the operator to misunderstand the true condition of the plant is also an unanalyzed condition and should be reported.

The Commission recognizes that the licensee may use engineering judgment and experience to determine whether an unanalyzed condition existed. It is not intended that this paragraph apply to minor variations in individual parameters, or to problems concerning single pieces of equipment. For example, at any time, one or more safety-related components may be out of service due to testing, maintenance, or a fault that has not yet been repaired. Any trivial single failure or minor error in performing surveillance tests could produce a situation in which two or more often unrelated, safety-grade components are out-of-service. Technically, this is an unanalyzed condition. However, these events should be reported only if they involve functionally related components or if they significantly compromise plant safety. When applying engineering judgement, and there is a doubt regarding whether to report or not, the Commission's policy is that licensees should make the report.

Finally, this paragraph also includes material (e.g., metallurgical or chemical) problems that cause abnormal degradation of the principal safety barriers (i.e., the fuel cladding, reactor coolant system pressure boundary, or

the containment). Examples of this type of situation include:

(a) Fuel cladding failures in the reactor, or in the storage pool, that exceed expected values, or that are unique or widespread, or that are caused by unexpected factors, and would involve a release of significant quantities of fission products.

(b) Cracks and breaks in the piping or reactor vessel (steel or prestressed concrete) or major components in the primary coolant circuit that have safety relevance (steam generators, reactor coolant pumps, valves, etc.).

(c) Significant welding or material defects in the primary coolant system.

(d) Serious temperature or pressure transients.

(e) Loss of relief and/or safety valve functions during operation.

(f) Loss of containment function or integrity including:

(i) Containment leakage rates exceeding the authorized limits.

(ii) Loss of containment isolation valve function during tests or operation.

(iii) Loss of main steam isolation valve function during test or operation, or

(iv) Loss of containment cooling capability.

Paragraph 50.72(b)(1)(iii), encompassing a portion of proposed 50.72(b)(2), was reworded to correspond to a similar provision of 10 CFR 50.73(a)(2)(iii). Making the requirements of 10 CFR 50.72 and 50.73 similar in language increases the clarity of these rules and minimizes confusion.

The paragraph has also been reworded to make it clear that it applies only to acts of nature (e.g., tornadoes) and external hazards (e.g., railroad tank car explosion). References to acts of sabotage have been removed, since these are covered by § 73.71. In addition, threats to personnel from internal hazards (e.g., radioactivity releases) that hamper personnel in the performance of necessary duties are now covered by paragraph 50.72(b)(1)(vi). This paragraph covers those events involving an actual threat to the plant from an external condition or natural phenomenon, and where the threat or damage challenges the ability of the plant to continue to operate in a safe manner (including the orderly shutdown and maintenance of shutdown conditions). The licensee should decide if a phenomenon or condition actually threatens the plant. For example, a minor brush fire in a remote area of the site that is quickly controlled by fire fighting personnel and, as a result, did not present a threat to the plant should not be reported. However, a major forest fire, large-scale flood, or major earthquake that presents a clear threat to the plant should be reported. As another example, an industrial or transportation accident which occurs near the site, creating a plant safety concern, should be reported.

Paragraph 50.72(b)(1)(iv), encompassing events previously classified as Unusual Events, requires the reporting of those events that result in either automatic or manual actuation of the ECCS or would have resulted in activation of the ECCS if some component had not failed or an operator action had not been taken.

For example, if a valid ECCS signal were generated by plant conditions, and the operator were to put all ECCS pumps in pull-to-lock, though no ECCS discharge occurred, the event would be reportable.

A "valid signal" refers to the actual plant conditions or parameters satisfying the requirements for ECCS initiation. Excluded from this reporting requirement would be those instances where instrument drift, spurious signals, human error, or other invalid signals caused actuation of the ECCS. However, such events may be reportable under other sections of the Commission's regulations based upon other details: in particular, paragraph 50.72(b)(2)(ii) requires a report within four hours if an Engineered Safety Feature (ESF) is actuated.

Experience with notifications made pursuant to § 50.72 has shown that events involving ECCS discharge to the vessel are generally more serious than ESF actuations without discharge to the vessel. Based on this experience, the Commission has made this reporting criterion a "One-Hour Report."

Paragraph 50.72(b)(1)(v), encompassing events previously classified as Unusual Events, covers those events that would impair a licensee's ability to deal with an accident or emergency. Notifying the NRC of these events may permit the NRC to take some compensating measures and to more completely assess the consequences of such a loss should it occur during an accident or emergency.

Examples of events that this criterion is intended to cover are those in which any of the following are not available:

1. Safety parameter display system (SPDS).

2. Emergency Response Facilities (ERF's).

3. Emergency communications facilities and equipment including the Emergency Notification system (ENS).

4. Public prompt Notification System including sirens.

5. Plant monitors necessary for accident assessment.

Paragraph 50.72(b)(1)(vi), encompassing some portions of the proposed §§ 50.72(b) (2) and (6), has been revised to add the phrase, "including fires, toxic gas releases, or radioactive releases." This addition covers the "evacuation" portion of paragraph 50.72(b)(6)(iii) of the proposed rule. This change in wording for the final rule was made in response to public comments discussed above.

While paragraph 50.72(b)(1)(iii) of the final rule primarily captures acts of nature, paragraph 50.72(b)(1)(vi) captures other events, particularly acts by personnel. The Commission believes this arrangement of the reporting criteria in the final rule lends itself to more precise interpretion and is consistent with those pubic comments that requested closer coordination between the reporting requirements in this rule and other portions of the Commission's regulations.

This provision requires reporting of events, particularly those caused by acts of personnel, which endanger the safety of the plant or interfere with personnel in performance of duties necessary for safe plant operations.

The licensee must exercise some judgment in reporting under this section. For example, a small fire on site that did not endanger any plant equipment and that did not and could not reasonably be expected to endanger the plant, is not reportable.

Paragraph 50.72(b)(1) of the proposed rule was split into § 50.72(b)(1)(ii) and § 50.72(b)(2)(i) in the final rule in order to permit some type of reports to be made within four hours instead of one hour because these reports have less safety significance. In terms of their combined effect, the overall intent and scope of these paragraphs have not changed from those in the proposed rule. Since the types of events intended to be captured by this reporting requirement are similar to § 50.72(b)(1)(ii), except that the reactor is shut down, the reader should refer to the explanation of § 50.72(b)(1)(ii) for more details on intent.

Paragraph 50.72(b)(2) Although the reporting criteria contained in the subparagraphs of § 50.72(b)(2) were in the proposed rule, in response to public comment the Commission established this "Non-Emergency" category for those events with slightly less urgency and less safety significance that may be reported within four hours instead of one hour.

The Commission wants to obtain such reports from personnel who were on shift at the time of the event, when this

is possible, because these personnel will have a better knowledge of the circumstances associated with the vent. Reports made within four hours of the event should make this possible while not imposing the more rigid one hour requirements.

The reporting requirement in paragraph 50.72(b)(2)(i) is similar to a requirement in § 50.73. Moreover, except for referring to a shutdown reactor, this reporting requirement is also similar to the "One-Hour Report" in § 50.72(b)(1)(ii). However this paragraph applies to a reactor in shutdown condition. Events within this requirement have less urgency and can be reported within four hours as a "Non-Emergency."

Paragraph 50.72(b)(2)(ii) (proposed 50.72(b)(5)) is made a "Non-Emergency" in response to public comment, because the Commission agrees that the covered events generally have slightly less urgency and safety significance than those events included in the "One-Hour Reports."

The intent and scope of this reporting requirement have not changed from the proposed rule. This paragraph is intended to capture events during which an ESF actuates, either manually or automatically, or fails to actuate. ESFs are provided to mitigate the consequences of the event; therefore, (1) they should work properly when called upon and (2) they should not be challenged unnecessarily. The Commission is interested both in events where an ESF was needed to mitigate the consequences of the event (whether or not the equipment performed properly) and events where an ESF operated unnecessarily.

"Actuation" of multichannel ESF Actuation Systems is defined as actuation of enough channels to complete the minimum actuation logic. Therefore, single channel actuations, whether caused by failures or otherwise, are not reportable if they do not complete the minimum actuation logic.

Operation of an ESF as part of a planned test or operational evolution need not be reported. However, if during the test or evolution the ESF actuates in a way that is not part of the planned procedure, that actuation should be reported. For example, if the normal reactor shutdown procedure requires that the control rods be inserted by a manual reactor trip, the reactor trip need not be reported. However, if conditions develop during the shutdown that require an automatic reactor trip, such a reactor trip should be reported. The fact that the safety nalysis assumes that an ESF will ctuate automatically during an event oes not eliminate the need to report at actuation. Actuations that need not e reported are those initiated for easons other then to mitigate the onsequences of an event (e.g., at the iscretion of the licensee as part of a lanned procedure).

Paragraph 50.72(b)(2)(iii) (proposed 0.72(b)(4)) has been revised and implified.

The words "any instance of personal rror, equipment failure, or discovery of lesign or-procedural inadequacies" that ppeared in the proposed rule have been eplaced by the words "event or ondition." This simplification in anguage is intended to clarify what was confusing phrase to many of those who commented on the proposed rule. Also in response to public comment, this eporting requirement is a "Non-Emergency" to be reported within four ours instead of within one hour.

This paragraph is based on the assumption that safety-related systems and structures are intended to mitigate the consequences of an accident. While paragraph 50.72(b)(2)(ii) applies to actual demands for actuation of an ESF, paragraph 50.72(b)(2)(iii) covers an event where a safety system could have failed to perform its intended function because of one or more personnel errors, including procedure violations; equipment failures; or design, analysis, fabrication, construction, or procedural deficiencies. The event should be reported regardless of the situation or condition that caused the structure or system to be unavailable.

This reporting requirement is similar to one contained in § 50.73, thus reflecting public comment identifying the need for closer coordination of reporting requirements between § 50.72 and § 50.73.

This paragraph includes those safety systems designed to mitigate the consequences of an accident (e.g., containment isolation, emergency filtration). Hence, minor operational events such as valve packing leaks, which could be considered a lack of control of radioactive material, should not be reported under this paragraph. System leaks or other similar events may, however, be reportable under other paragraphs.

This paragraph does not include those cases where a system or component is removed from service as part of a planned evolution, in accordance with an approved procedure, and in accordance with the plant's Technical Specifications. For example, if the licensee removes part of a system from

service to perform maintenance, and the Technical Specifications permit the resulting configuration, and the system or component is returned to service within the time limit specified in the Technical Specifications, the action need not be reported under this paragraph. However, if, while the component is out of service, the licensee identifies a condition that could have prevented the system from performing its intended function (e.g., the licensee finds a set of relays that is wired incorrectly), that condition must be reported.

It should be noted that there are a limited number of single-train systems that perform safety functions (e.g., the High Pressure Coolant Injection System in BWRs). For such systems, loss of the single train would prevent the fulfillment of the safety function of that system and, therefore, must be reported even though the plant Technical Specifications may allow such a condition to exist for a specified length of time. Also, if a potentially serious human error is made that could have prevented fulfillment of a safety function, but recovery factors resulted in the error being corrected, the error is still reportable.

The Commission recognizes that the application of this and other paragraphs of this section involves a technical judgment by licensees. In this case, a technical judgment must be made whether a failure or operator action that disabled one train of a safety system could have, but did not, affect a redundant train. If so, this would constitute an event that "could have prevented" the fulfillment of a safety function, and, accordingly, must be reported.

If a component fails by an apparently random mechanism, it may or may not be reportable if the functionally redundant component could fail by the same mechanism. To be reportable, it is necessary that the failure constitute a condition where there is reasonable doubt that the functionally redundant train or channel would remain operational until it completed its safety function or is repaired. For example, if a pump fails because of improper lubrication, there is a reasonable expectation that the functionally redundant pump, which was also improperly lubricated, would have also failed before it completed its safety function, then the failure is reportable and the potential failure of the functionally redundant pump must be reported.

Interaction between systems, particularly a safety system and a non-safety system, is also included in this

criterion. For example, the Commission is increasingly concerned about the effect of a loss or degradation of what had been assumed to be nonessential inputs to safety systems. Therefore, this paragraph also includes those cases where a service (e.g., heating, ventilation, and cooling) or input (e.g., compressed air) which is necessary for reliable or long-term operation of a safety system is lost or degraded. Such loss or degradation is reportable, if the proper fulfillment of the safety function is not or can not be assured. Failures that affect inputs or services to systems that have no safety function need not be reported.

Finally, the Commission recognizes that the licensee has to decide when personnel actions *could* have prevented fulfillment of a safety function. For example, when an individual improperly operates or maintains a component, that person might conceivably have made the same error for all of the functionally redundant components (e.g., if an individual incorrectly calibrates one bistable amplifier in the Reactor Protection System, that person could conceivably incorrectly calibrate all bistable amplifiers). However, for an event to be reportable it is necessary that the actions actually affect or involve components in more than one train or channel of a safety system, and the result of the actions must be undersirable from the perspective of protecting the health and safety of the public. The components can be functionally redundant (e.g., two pumps in different trains) or not functionally redundant (e.g., the operator correctly stops a pump in Train "A" and instead of shutting the pump discharge valve in Train "A," he mistakenly shuts the pump discharge valve in Train "B").

Paragraphs 50.72(b)(2)(iv) (proposed 50.72(b)(6)) has been changed to clarify the requirement to report releases of radioactive material. The paragraph is similar to § 20.403 but places a lower threshold for reporting events at commercial power reactors. The lower threshold is based on the significance of the breakdown of the licensee's program necessary to have a release of this size, rather than on the significance of the impact of the actual release. The existing licensee radioactive material effluent release monitoring programs and their associated assessment capabilities are sufficient to satisfy the intent of 50.72(b)(2)(iv).

Based upon public comment and a reevaluation by the Commission staff, the reporting threshold has been changed from "25%" in the proposed rule to "2 times" in the final rule and has been reclassified as a "Non-Emergency" to be reported within four hours instead of within 1 hour:

Also this reporting requirement has been changed to make a more uniform requirement by referring to specific release criteria instead of referring only to Technical Specifications that may vary somewhat among facilities.

This reporting requirement is intended to capture those events that may lead to an accident situation where significant amounts of radioactive material could be released from the facility. Unplanned releases should occur infrequently; however, if they occur at the levels specified, at least moderate defects have occurred in the safety design or operational control established to avoid their occurrence and, therefore, such events should be reported.

Normal operating limits for radioactive effluent releases are based on the limits of 10 CFR Part 20 which establishes maximum annual average concentration in unrestricted areas. This reporting requirement addresses concentrations averaged over a one hour period and represents less than 0.1% of the annual quantities of radioactive materials permitted to be released by 10 CFR Part 20.

Paragraph 50.72(b)(2)(v) (proposed rule 50.72(b)(7)) has three changes. The first eliminates the phrase "occurring onsite" because it is implied by the scope of the rule. The second replaces "injury involving radiation" with "radioactively contaminated person." This change was made because of the difficulty in defining injury due to radiation. and more importantly, because 10 CFR Part 20 captures events involving radiation exposure.

The third change, in response to public comment, was to make this reporting requirement a four-hour notification, instead of one-hour notification. This change was made because these events have slightly less safety significance than those required to be reported within one hour.

Paragraph 50.72(b)(2)(vi) (not in proposed rule) besides covering some events such as release of radioactively contaminated tools or equipment to the public that may warrent NRC attention, also covers those events that would not otherwise warrant NRC attention except for the interest of the news media, other government agencies, or the public. In terms of its effect on licensees. this is not a new reporting requirement because the threshold for reporting injuries and radioactive release was much lower under the proposed rule. This criterion will capture those events previously reported under other criteria when such events require the NRC to

respond because of media or public attention.

Paragraph 50.72(c) (proposed 50.72(c)) has remained essentially unchanged from the proposed rule. except for addition of the title "Followup Notification" and some renumbering.

This paragraph is intended to provide the NRC with timely notification when an event becomes more serious or additional information or new analyses clarify an event.

This paragraph also permits the NRC to maintain a continuous communications channel because of the need for continuing follow-up information or because of telecommunications problems.

IV. Regulatory Analysis

The Commission has prepared a regulatory analysis on this regulation. The analysis examines the costs and benefits of the Rule as considered by the Commission. A copy of the regulatory analysis is available for inspection and copying for a fee at the NRC Public Document Room, 1717 H Street, NW., Washington, D.C. Single copies of the analysis may be obtained from Eric W. Weiss, Office of Inspection and . Enforcement, U.S. Nuclear Regulatory Commission, Washington, D.C. 20555, Telephone (301) 492–4973.

V. Paperwork Reduction Act Statement

The information collection requirements contained in this final rule have been approved by the Office of Management and Budget pursuant to the Paperwork Reduction Act, Pub. L. 96–511 (clearance number 3150–0011).

VI. Regulatory Flexibility Certification

In accordance with the Regulatory Flexibility Act of 1980, 5 U.S.C. 605(b). the Commission hereby certifies that this regulation will not have a significant economic impact on a substantial number of small entities. This final rule affects electric utilities that are dominant in their respective service areas and that own and operate nuclear utilization facilities licensed under sections 103 and 104b. of the Atomic Energy Act of 1954, as amended. The amendments clarify and modify presently existing notification requirements. Accordingly, there is no new, significant economic impact on these licensees, nor do the affected licensees fall within the scope of the definition of "small entities" set forth in the Regulatory Flexibility Act or within the Small Business Size Standards set forth in regulations issued by the Small Business Administration at 13 CFR Part 121.

List of Subjects in 10 CFR Part 50

Antitrust. Classified information, Fire prevention. Incorporation by reference. Intergovernmental relations, Nuclear power plants and reactors. Penalty, Radiation protection, Reactor siting criteria. Reporting and recordkeeping requirements.

Pursuant to the Atomic Energy Act of 1954. as amended, the Energy Reorganization Act of 1974, as amended. and section 552 and 553 of Title 5 of the United States Code, the following amendments to Title 10, Chapter I, Code of Federal Regulations, Part 50 are published as a document subject to codification.

PART 50—DOMESTIC LICENSING OF PRODUCTION AND UTILIZATION FACILITIES

1. The authority citation for Part 50 continues to read as follows:

Authority: Secs. 103, 104, 161, 182, 183, 186, 189, 68 Stat. 936, 937, 948, 953, 954, 955, 956, as amended, sec. 234, 83 Stat 1244, as amended (42 U.S.C. 2133, 2134, 2201, 2232, 2233, 2236, 2239, 2282); secs. 201, 202, 206, 68 Stat. 1242, 1244, 1246, as amended (42 U.S.C. 5841, 5842, 5846), unless otherwise noted.

Section 50.7 also issued under Pub. L. 95–601, sec. 10, 92 Stat. 2951 (42 U.S.C. 5851). Sections 50.58, 50.91 and 50.92 also issued under Pub. L. 97–415, 96 Stat. 2073 (42 U.S.C. 2239). Section 50.78 also issued under sec. 122, 68 Stat. 939 (42 U.S.C. 2152). Sections 50.80–50.81 also issued under sec. 184, 68 Stat. 954, as amended (42 U.S.C. 2234). Sections 50.100–50.102 also issued under sec. 186, 68 Stat. 955 (42 U.S.C. 2236).

For the purposes of sec. 223, 68 Stat. 958, as amended (42 U.S.C. 2273), §§ 50.10 (a), (b), and (c), 50.44, 50.46, 50.48, 50.54, and 50.80(a) are issued under sec. 161b, 68 Stat. 948, as amended (42 U.S.C. 2201(b)); §§ 50.10 (b) and (c) and 50.54 are issued under sec. 161i, 68 Stat. 949, as amended (42 U.S.C. 2201(i)); and §§ 50.55(e), 50.59(b), 50.70, 50.71, 50.72, and 50.78 are issued under sec. 161o, 68 Stat. 950, as amended (42 U.S.C. 2201(o)).

2. A new paragraph (z) is added to § 50.54 to read as follows:

§ 50.54   Conditions of licenses.

*   *   *   *   *

(z) Each licensee with a utilization facility licensed pursuant to sections 103 or 104b. of the Act shall immediately notify the NRC Operations Center of the occurrence of any event specified in § 50.72 of this part.

3. Section 50.72 is revised to read as follows:

§ 50.72 Immediate notification requirements for operating nuclear power reactors.

(a) *General Requirements.*[1] (1) Each nuclear power reactor under § 50.21(b) or § 50.22 of this part shall notify the NRC Operations Center via the Emergency Notification System of:

(i) The declaration of any of the Emergency Classes specified in the licensee's approved Emergency Plan;[2] or

(ii) Of those non-Emergency events specified in paragraph (b) of ths section.

(2) If the Emergency Notification System is inoperative, the licensee shall make the required notifications via commerical telephone service, other dedicated telephone system, or any other method which will ensure that a report is made as soon as practical to the NRC Operations Center.[3]

(3) The licensee shall notify the NRC immediately after notification of the appropriate State or local agencies and not later than one hour after the time the licensee declares one of the Emergency Classes.

(4) When making a report under paragraph (a)(3) of this section, the licensee shall identify:

(i) The Emergency Class declared; or

(ii) Either paragraph (b)(1), "One-Hour Report," or paragraph (b)(2), "Four-Hour Report," as the paragraph of this section requiring notification of the Non-Emergency Event.

(b) *Non-Emergency Events.* (1) *One-Hour Reports.* If not reported as a declaration of an Emergency Class under paragraph (a) of this section, the licensee shall notify the NRC as soon as practical and in all cases within one hour of the occurrence of any of the following:

(i)(A) The initiation of any nuclear plant shutdown required by the plant's Technical Specifications.

(B) Any deviation from the plant's Technical Specifications authorized pursuant to § 50.54(x) of this part.

(ii) Any event or condition during operation that results in the condition of the nuclear powerplant, including its principal safety barriers, being seriously degraded; or results in the nuclear powerplant being:

(A) In a unanalyzed condition that significantly compromises plant safety;

(B) In a condition that is outside the design basis of the plant; or

(C) In a condition not covered by the plant's operating and emergency procedures.

(iii) Any natural phenomenon or other external condition that poses an actual threat to the safety of the nuclear power-plant or significantly hampers site personnel in the performance of duties necessary for the safe operation of the plant.

(iv) Any event that results or should have resulted in Emergency Core Cooling System (ECCS) discharge into the reactor coolant system as a result of a valid signal.

(v) Any event that results in a major loss of emergency assessment capability, offsite response capability, or communications capability (e.g., significant portion of control room indication, Emergency Notification System, or offsite notification system).

(vi) Any event that poses an actual threat to the safety of the nuclear powerplant or significantly hampers site personnel in the performance of duties necessary for the safe operation of the nulcear powerplant including fires, toxic gas releases, or radioactive releases.

(2) *Four-Hour Reports.* If not reported under paragraphs (a) or (b)(1) of this section, the licensee shall notify the NRC as soon as practical and in all cases, within four hours of the occurrence of any of the following:

(i) Any event, found while the reactor is shutdown, that, had it been found while the reactor was in operation, would have resulted in the nuclear powerplant, including its principal safety barriers, being seriously degraded or being in an unanalyzed condition that significantly compromises plant safety.

(ii) Any event or condition that results in manual or automatic actuation of an Engineered Safety Feature (ESF), including the Reactor Protection System (RPS). However, actuation of an ESF, including the RPS, that results from and is part of the preplanned sequence during testing or reactor operation need not be reported.

(iii) Any event or condition that alone could have prevented the fulfillment of the safety function of structures or systems that are needed to:

(A) Shut down the reactor and maintain it in a safe shutdown condition,

(B) Remove residual heat,

(C) Control the release of radioactive material, or

(D) Mitigate the consequences of an accident.

(iv)(A) Any airborne radioactive release that exceeds 2 times the applicable concentrations of the limits specified in Appendix B, Table II of Part 20 of this chapter in unrestricted areas,

when averaged over a time period of one hour.

(B) Any liquid effluent release that exceeds 2 times the limiting combined Maximum Permissible Concentration (MPC) (see Note 1 of Appendix B to Part 20 of this chapter) at the point of entry into the receiving water (i.e., unrestricted area) for all radionuclides except tritium and dissolved noble gases, when averaged over a time period of one hour. (Immediate notifications made under this paragraph also satisfy the requirements of paragraphs (a)(2) and (b)(2) of § 20.403 of Part 20 of this chapter.)

(v) Any event requiring the transport of a radioactively contaminated person to an offsite medical facility for treatment.

(vi) Any event or situation, related to the health and safety of the public or onsite personnel, or protection of the environment, for which a news release is planned or notification to other government agencies has been or will be made. Such an event may include an onsite fatality or inadvertent release of radioactively contaminated materials.

(c) *Followup Notification.* With respect to the telephone notifications made under paragraphs (a) and (b) of this section, in addition to making the required initial notification, each licensee, shall during the course of the event:

(1) *Immediately report:* (i) any further degradation in the level of safety of the plant or other worsening plant conditions, including those that require the declaration of any of the Emergency Classes, if such a declaration has not been previously made, or (ii) any change from one Emergency Class to another, or (iii) a termination of the Emergency Class.

(2) *Immediately report:* (i) the results of ensuing evaluations or assessments of plant conditions, (ii) the effectiveness of response or protective measures taken, and (iii) information related to plant behavior that is not understood.

(3) Maintain an open, continuous communication channel with the NRC Operations Center upon request by the NRC.

Dated: at Washington, D.C. this 23d day of August, 1983.

For the Nuclear Regulatory Commission.

Samuel J. Chilk,

Secretary of the Commission.

[FR Doc. 83-23602 Filed 8-26-83; 8:45 am]

BILLING CODE 7590-01-M

---

[1] Other requirements for immediate notification of the NRC by licensed operating nuclear power reactors are contained elsewhere in this chapter, in particular, § 20.205, § 20.403, § 50.36, and § 73.71.

[2] These Emergency Classes are addressed in appendix E of this part.

[3] Commercial telephone number of the NRC Operations Center is (202) 951-0550.

APPENDIX E

10 CFR 50.73 INCLUDING
STATEMENT OF CONSIDERATIONS

Published in the *Federal Register*
on July 26, 1983
(Vol. 48, No. 144, pages 33850-33860)

NOTE: This *Federal Register* notice does not provide a current and correct version of 10
CFR 50.73, which has been amended several times since 1983. Its purpose
here is to present the Statement of Considerations, which explains the basic
reporting requirements of 10 CFR 50.73.

NUCLEAR REGULATORY COMMISSION

10 CFR Parts 20 and 50

Licensee Event Report System

AGENCY: Nuclear Regulatory Commission.

ACTION: Final rule.

SUMMARY: The Commission is amending its regulations to require the reporting of operational experience at nuclear power plants by establishing the Licensee Event Report (LER) system. The final rule is needed to codify the LER reporting requirements in order to establish a single set of requirements that apply to all operating nuclear power plants. The final rule applies only to licensees of commercial nuclear power plants. The final rule will change the requirements that define the events and situations that must be reported, and will define the information that must be provided in each report.

EFFECTIVE DATE: January 1, 1984. The incorporation by reference of certain publications listed in the regulations is approved by the Director of the Federal Register as of January 1, 1984.

FOR FURTHER INFORMATION CONTACT: Frederick J. Hebdon, Chief, Program Technology Branch, Office for Analysis and Evaluation of Operational Data, U.S. Nuclear Regulatory Commission, Washington, D.C. 20555; Telephone (301) 492-4480.

SUPPLEMENTARY INFORMATION:

I. Background

On May 6, 1982, the NRC published in the Federal Register (47 FR 19543)[1] a Notice of Proposed Rulemaking that would modify and codify the existing Licensee Event Report (LER) system. Interested persons were invited to submit written comments to the Secretary of the Commission by July 6, 1982. Numerous comments were received. After consideration of the comments and other factors involved, the Commission has amended the proposed requirements published for public comment by clarifying the scope and content of the requirements, particularly the criteria that define which operational events must be reported.

The majority of the comments on the proposed rule: (1) Questioned the meaning and intent of the criteria that defined the events which must be reported, (2) questioned the need for reporting certain specific types of events, and (3) questioned the need for certain information that would be required to be included in an LER. Section III of this notice discusses the comments in more detail.

[1] Copies of the documents are available for public inspection and copying for a fee at the Public Document Room at 1717 H Street NW, Washington, D.C.

II. Rulemaking initiation

The Nuclear Plant Reliability Data (NPRD) system is a voluntary program for the reporting of reliability data by nuclear power plant licensees. On January 30, 1980 (45 FR 6793),[1] the NRC published an Advance Notice of Proposed Rulemaking that described the NPRD system and invited public comment on an NRC plan to make it mandatory. Forty-four letters were received in response to the advanced notice. These comments generally opposed making the NPRD system mandatory on the grounds that reporting of reliability data should not be made a regulatory requirement.

In December 1980, the Commission decided that the requirements for reporting of operational experience data needed major revision and approved the development of an Integrated Operational Experience Reporting (IOER) system. The IOER system would have combined, modified, and made mandatory the existing Licensee Event Report (LER) system and the NPRD system. SECY 80-507[1] discusses the IOER system.

As a result of the Commission's approval of the concept of an IOER system, the NRC published another advance notice on January 15, 1981 (46 FR 3541). This advance notice explained why the NRC needed operational experience data and described the deficiencies in the existing LER and NPRD systems.

On June 6, 1981, the Institute of Nuclear Power Operations (INPO) announced that because of its role as an active user of NPRDs data it would assume responsibility for management and funding of the NPRD system. Further, INPO decided to develop criteria that would be used in its management audits of member utilities to assess the adequacy of participation in the NPRD system.

The two principal deficiencies that had previously made the NPRD system an inadequate source of reliability data were the inability of its committee management structure to provide the necessary technical direction and a low level of participation by the utilities. The commitments and actions by INPO provided a basis for confidence that these two deficiencies would be corrected. For example, centralizing the management and funding of NPRDS within INPO should overcome the previous difficulties associated with management by a committee and funding from several independent organizations. Further, with INPO focusing upon a utility's participation in

NPRDS as a specific evaluation parameter during routine management and plant audit activities, the level of utility participation, and therefore, the quality and quantity of NPRDS data, should significantly increase. However, the Commission will continue to have an active role in NPRDS by participating in an NPRDS User's Group, by periodically assessing the quality and quantity of information available from NPRDS, and by auditing the timely availability of the information to the NRC.

Since there was a likelihood that NPRDS under INPO direction would meet the NRC's need for reliability data, it was no longer necessary to proceed with the IOERS. Hence, the collection of detailed technical descriptions of significant events could be addressed in a separate rulemaking to modify and codify the existing LER reporting requirements. See SECY 81–494 for additional details concerning IOERS.

However, the Commission wishes to make it explicitly clear that it is relaxing the reporting requirements with the expectation that sufficient utility participation, cooperation, and support of the NPRD system will be forthcoming. If the NPRD system does not become operational at a satisfactory level in a reasonable time, remedial action by the Commission in the form of additional rulemaking may become necessary.

On October 6, 1981, the NRC published an advanced notice (46 FR 49134) that deferred development of the IOER system and sought public comment on the scope and content of the LER system. Six comment letters were received in response to this ANPRM. All of the comments received were reviewed by the staff and were considered in the development of the proposed LER rule. See SECY 82–3 [1] for additional details.

This rule identifies the types of reactor events and problems that are believed to be significant and useful to the NRC in its effort to identify and resolve threats to public safety. It is designed to provide the information necessary for engineering studies of operational anomalies and trends and patterns analysis of operational occurrences. The same information can also be used for other analytic procedures that will aid in identifying accident precursors.

The Commission believes that the NRC should continue to seek an improved operational data system that will maximize the value of operational data. The system should encompass and integrate operational data of events and problem sequences identified in this rule, NPRDS data, and such other information as is required for a comprehensive integrated analytically-versatile system.

The Brookhaven Study, published as BNL/NUREG 51609, NUREG/CR 3206, discusses data collection and storage procedures to support multivariate, multicase analysis. While the range of reactor configurations in the U.S. nuclear industry presents some methodological and interpretative problems, these difficulties should not be insurmountable. The Commission believes that the NRC should have as a specific objective the development, demonstration, and implementation of an integrated system for collecting and analyzing operational data that will employ the predictive and analytical potential of multicase, multivariate analyses. Accordingly, the staff has been directed to undertake the work necessary to develop and demonstrate such a cost-effective integrated system of operational data collection and analyses.

If the design of the system demonstrates that such a system is feasible and cost-effective, development of the system to the point of initiating rule should be completed by July 1986.

III. Analysis of Comments

The Commission received forty-seven (47) letters commenting on the proposed rule. Copies of those letters and a detailed analysis of the comments are available for public inspection and copying for a fee at the NRC Public Document Room at 1717 H Street, NW., Washington, D.C. A number of the more substantive issues are discussed below.

Licensee Resources

Of particular concern to the Commission was the impact that the proposed rule would have on the resources used by licensees to prepare LERs. The Commission's goal was to assure that the scope of the rule would not increase the overall level of effort above that currently required to comply with the existing LER requirements. Thirty letters of the 47 received contained comments on the overall acceptability of the proposed rule or commented directly on the question of scope and/or resources associated with the proposed rule. The views of the commenters can be characterized as follows:

1. Five commenters felt that the scope and level of effort would be greatly expanded by the proposed rule. Estimates included an increase of 100 man-years for the entire industry, an increase of three times the current effort, and an increase of $100,000 and 2 man-years annually for each plant.

2. Four commenters felt that the level of effort would be increased but not significantly.

3. One commenter felt that the proposed rule would have a minimal effect on the level of effort required.

4. Two commenters felt that the proposed rule would significantly reduce the number of LERs filed.

5. Thirteen commenters endorsed the objective of improving LER reporting but felt that changes in the proposed rule were needed. These commenters did not directly address the resource issue.

6. Five commenters endorsed the proposed rule and/or felt that it was a significant improvement over the existing reporting requirements.

Based on these comments and its own assessment of the impact of this rule, the Commission has concluded that the impact of this rule will be no greater than the impact of the existing LER requirements, and this rule will not place an unacceptable burden on the affected licensees.

Relationship Between the LER Rule (§ 50.73) and the Immediate Notification Rule (§ 50.72)

As a parallel activity to the preparation of § 50.73, the Commission is amending its regulations (§ 50.72) which require that licensees for nuclear power plants notify the NRC Operations Center of significant events that occur at their plants. On December 21, 1981, the Commission published in the Federal Register a proposed rule (46 FR 61894) that described the planned changes in § 50.72.

The Federal Register notice accompanying the proposed LER rule (i.e., § 50.73) stated that additional changes anticipated to § 50.72 would be made but they would be "* * * largely administrative and the revised § 50.72 would not be significantly modified nor would it be published again for public comment." Several commenters disagreed with this conclusion.

The commenters did, however, agree with the Commission's position that inconsistencies and overlapping requirements between the two rules need to be eliminated.

The Commission has carefully reviewed the proposed requirements in the LER and Immediate Notification rules and has concluded that although changes to both have been made (largely in response to public comments) to clarify the intent of the rules, the original intent and scope have not been significantly changed. Therefore, the Commission has concluded that these two rules need not be published again for public comment.

Engineering Judgment

In the Federal Register notice that accompanied the proposed rule, the Commission stated that licensee's engineering judgment may be used to decide if an event is reportable. Several commenters expressed the belief that some wording should be added to the rule of reflect that the NRC will also use judgment in enforcement of this regulation where the licensee is requested to use engineering judgment.

The Commission believes that the LER rule adequately discusses the need for and application of the concept of "engineering judgment." The concept itself includes the recognition of the existence of a reasonable range of interpretation regarding this rule, and consequently the Commission recognizes and hereby acknowledges the need for flexibility in enforcement actions associated with this rule. The Commission believes that this concept is sufficiently clear and that additional explicit guidance is not necessary.

Reporting Schedule

In the Federal Register notice that accompanied the proposed rule, the Commission stated that it had not yet decided if the reports should be submitted in fifteen days or thirty days following discovery of a reportable event. Many commenters stated that the time frame for reporting LERs should not be less than thirty days after the discovery of a reportable event.

One commenter estimated the impact of a requirement to submit a report sooner than 30 days following discovery of a reportable event would be an increase of approximately 40 man years per year for the currently operating plants. In addition the commenter estimated that if a summary report were also required the reporting burden would increase an additional 12 man years for the currently operating plants.

In response to these comments, the Commission has decided to require that LERs be submitted within 30 days of discovery of a reportable event or situation.

Reporting of Reactor Trips

Section 50.73(a)(1) of the proposed rule (§ 50.73(a)(2)(iv) of the final rule) required reporting of any event which results in an unplanned manual or automatic actuation of any Engineered Safety Feature (ESF) including the Reactor Protection System (RPS). Many commenters agreed that these events should be trended and analyzed, but disagreed that they deserve to be singled out as events of special significance (i.e., events reportable as

LERs). They noted that reports of RPS actuations are already reported to the NRC in the Monthly Operating Status Report, as well as telephoned to the NRC Operations Center.

In addition, the Institute of Nuclear Power Operations (INPO) analyzed the frequency of reactor scrams during a one-month period. This analysis indicated that an average of 55 reactor trips would be reportable each month under the proposed rule. INPO equated this to 660 additional LERs per year for all currently operating plants, or approximately 32 man-years of additional effort for all the currently operating plants based upon the assumption that each LER requires 100 man-hours of effort to prepare and analyze.

The Commission still believes that ESF actuations, including reactor trips, frequently are associated with significant plant transients and are indicative of events that are of safety significance. In addition, if the ESFs are being challenged during routine transients, that fact is of safety significance and should be reported.

In addition, the Commission does not agree with the estimate that each LER submitted for a routine reactor trip would require, on the average, 100 man-hours to prepare and analyze. Licensees are already required to make internal evaluation of and document significant events, including reactor trips. Therefore, the incremental impact of preparing and analyzing the LER should be significantly less than 100-man hours. In addition, the actual increase in burden would be offset by reductions in the burden of reporting less significant events that would no longer be reportable.

Coordination With Other Reporting Requirements

Several commenters noted that the proposed rule did not appear to be coordinated with other existing reporting requirements, and that duplication of licensee effort might result. They recommended that LER reporting be consolidated to eliminate potential duplication of other existing reporting requirements.

The Commission has reviewed existing NRC reporting requirements (e.g., 10 CFR Parts 20 and 21, § 50.55(e), § 50.72, § 50.73, § 73.71, and NUREG–0654) and has attempted, to the extent practicable, to eliminate redundant reporting and to ensure that the various reporting requirements are consistent. Many of the changes in the final LER rule are as a result of this effort. These changes resulted in extensive revisions in the wording of criteria contained in

this rule, but did not change the original scope of intent of the requirements. In addition, in order to make the requirements in §§ 50.72 and 50.73 more compatible, the order (i.e., numbering) of the criteria in § 50.73 has been changed. The changes are noted in the discussion of each paragraph below.

Finally, conforming amendments are being made to various sections of Parts 20 and 50 in order to reduce the redundancy in reporting requirements that apply to operating nuclear power plants. In general, these amendments will require that:

1. Licensees that have an Emergency Notification System (ENS) make the reports required by the subject sections via the ENS. All other licensees will continue to make the reports to the Administrator of the appropriate NRC Regional Office.

2. Written reports required by the subject sections be submitted to the NRC Document Control Desk in Washington, D.C., with a copy to the appropriate Regional Offices.

3. Holders of licenses to operate a nuclear power plant submit the written reports required by the subject sections in accordance with the procedures described in § 50.73(b).

The criteria contained in the subject sections which define a reportable event have not been modified.

Similar changes are also planned as part of current activities to make more substantive changes to Part 21, § 50.55(e), and § 73.71.

Nonconservative Interdependence

Several commenters expressed difficulty in understanding the meaning of the phrase "nonconservative interdependence" as used in the proposed § 50.73(a)(3). The wording of § 50.73(a)(3) (§ 50.73(a)(2)(vii) of this final rule) has been changed to eliminate the phrase "non conservative interdependence" by specifically defining the types of events that should be reported. The revised paragraph does not, however, change the intent of the original paragraph.

Sabotage and Threats of Violence

Several commenters noted that the security-related reporting requirements of § 50.73(a)(6) (§ 50.73(a)(2)(iii) of this final rule)) were already contained in greater detail in 10 CFR 73.71. For instance, § 73.71 requires an act of sabotage to be reported immediately, followed by a written report within 15 days. The proposed rule would have required an LER to be filed within 30 days. Although distribution of reports is somewhat different, redundant reporting would have occurred. The commenters recommended that the Commission ensure consistency between §§ 50.73 and 73.71.

In response to these comments the Commission has deleted the reporting of sabotage and threats of violence from § 50.73 because these situations are adequately covered by the reporting requirements contained in § 73.71.

Evacuation of Rooms or Buildings

Many commenters stated that the reporting of in-plant releases of radioactivity that require evacuation of individual rooms (§ 50.73(a)(7) in the proposed rule or (§ 50.73(a)(2)(x) of this final rule) was inconsistent with the general thrust of the rule to require reporting of significant events. They noted that minor spills, small gaseous waste releases, or the disturbance of contaminated particulate matter (e.g., dust) may all require the temporary evacuation of individual rooms until the airborne concentrations decrease or until respiratory protection devices are utilized. They noted that these events are fairly common and should not be reportable unless the required evacuation affects the entire facility or a major portion thereof.

In response to these comments the wording of this criterion (§ 50.73(a)(2)(x) in the final rule) has been changed to significantly narrow the scope of the criterion to include only those events which significantly hamper the ability of site personnel to perform safety-related activities (e.g., evacuation of the main control room).

Energy Industry Identification System

Many commenters noted that the requirement to report the Energy Industry Identification System (EIIS) component function identifier and system name of each component or system referred to in the LER description would be a significant burden on the licensee.

They suggested instead that the NPRDS component identifiers be used in place of the EIIS component identifiers which are not yet widely used by the industry.

The Commission continues to believe that EIIS system names and component function identifiers are needed in order that LERs from different plants can be compared. We do not, however, suggest that the EIIS identifiers be used throughout the plant, but only that they be added to the LER as it is written. A simple, inexpensive table could be used to translate plant identifiers into equivalent EIIS identifiers.

The Commission considered the system and component identifiers used

in NPRDS as an alternative. It is our understanding, however, the NPRDS will soon adopt the EIIS system titles, so a distinction should no longer exist. In addition, LERs frequently include systems that are not included in the scope of NPRDS (i.e., an NPRDS system identification does not exist) while EIIS, on the other hand, includes all of the systems commonly found in commercial nuclear power plants. Further, NPRDS includes only 39 component identifiers (e.g., valve, pump). The Commission believes that this limited number does not provide a sufficiently detailed description of the component function involved.

Function of Failed Components and Status of Redundant Components

Many commenters said that information required in (§ 50.73(b)(2) (vi) and (vii) of the proposed rule should not be a requirement in the LER. They argued that this information is readily available in documents previously submitted to the NRC by licensees and are available for reference.

The final rule (§ 50.73(b)(2)(i)(G)) has been modified to narrow the scope of the information requested by the Commission.

While this general information may be available in licensee documents previously submitted to the NRC, the Commission believes that a general understanding of the event and its significance should be possible without reference to additional documentation which may not be readily or widely available, particularly to the public.

The Commission continues to believe that the licensee should prepare an LER in sufficient depth so that knowledgeable readers who are conversant with the design of commercial nuclear power plants, but are not familiar with the details of a particular plant, can understand the general characteristics of the event (e.g., the cause, the significance, the corrective action). As suggested by the commenters, more detailed information to support engineering evaluations and case studies will be obtained, as needed, directly from the previously submitted licensee documents.

Engineering Evaluations

The overview discussion of the proposed rule contains the following statement: "If the NRC staff decides that the event was especially significant from the standpoint of safety, the staff may request that the licensee perform an engineering evaluation of the event and describe the results of that evaluation."

Several commenters argued that the inclusion of the requirement that the licensee perform an engineering evaluation of certain events at the staff's request appeared unjustified and would add substantially to the burden of reporting. They argued that the licensee should be required to submit only the specific additional information required for the necessary engineering evaluation rather than to perform the evaluation.

The rule has been modified to require only the submittal of any necessary additional information requested by the Commission in writing.

IV. Specific Findings

Overview of the LER System

When this final LER rule becomes effective, the LER will be a detailed narrative description of potentially significant safety events. By describing in detail the event and the planned corrective action, it will provide the basis for the careful study of events or conditions that might lead to serious accidents. If the NRC staff decides that the event was especially significant from the standpoint of safety, the staff may request that the licensee provide additional information and data associated with the event.

The licensee will prepare an LER for those events or conditions that meet one or more of the criteria contained in § 50.73(a). The criteria are based primarily on the nature, course, and consequences of the event. Therefore, the final LER rule requires that events which meet the criteria are to be reported regardless of the plant operating mode or power level, and regardless of the safety significance of the components, systems, or structures involved. In trying to develop criteria for the identification of events reportable as LERs, the Commission has concentrated on the potential consequences of the event as the measure of significance. Therefore, the reporting criteria, in general, do not specifically address classes of initiating events or causes of the event. For example, there is no requirement that all personnel errors be reported. However, many reportable events will involve or have been initiated by personnel errors.

Finally, it should be noted that licensees are permitted and encouraged to report any event that does not meet the criteria contained in § 50.73(a), if the licensee believes that the event might be of safety significance, or of generic interest or concern. Reporting requirements aside, assurance of safe operation of all plants depends on accurate and complete reporting by each licensee of all events having potential safety significance.

Paragraph-by-Paragraph Explanation of the LER Rule

The significant provisions of the final LER rule are explained below. The explanation follows the order in the proposed rule.

Paragraph 50.73(a)(2)(iv) (proposed paragraph 50.73(a)(1)) requires reporting of: "Any event or condition that resulted in manual or automatic actuation of any Engineered Safety Feature (ESF), including the Reactor Protection System (RPS). However, actuation of an ESF, including the RPS, that resulted from and was part of the preplanned sequence during testing or reactor operation need not be reported.

This paragraph requires events to be reported whenever an ESF actuates either manually or automatically, regardless of plant status. It is based on the premise that the ESFs are provided to mitigate the consequences of a significant event and, therefore: (1) They should work properly when called upon, and (2) they should not be challenged frequently or unnecessarily. The Commission is interested both in events where an ESF was needed to mitigate the consequences (whether or not the equipment performed properly) and events where an ESF operated unnecessarily.

"Actuation" of multichannel ESF Actuation Systems is defined as actuation of enough channels to complete the minimum actuation logic (i.e., activation of sufficient channels to cause activation of the ESF Actuation System). Therefore, single channel actuations, whether caused by failures or otherwise, are not reportable if they do not complete the minimum actuation logic.

Operation of an ESF as part of a planned operational procedure or test (e.g., startup testing) need not be reported. However, if during the planned operating procedure or test, the ESF actuates in a way that is not part of the planned procedure, that actuation must be reported. For example, if the normal reactor shutdown procedure requires that the control rods be inserted by a manual reactor trip, the reactor trip need not be reported. However, if conditions develop during the shutdown that require an automatic reactor trip, such a reactor trip must be reported.

The fact that the safety analysis assumes that an ESF will actuate automatically during certain plant conditions does not eliminate the need to report that actuation. Actuations that need not be reported are those initiated for reasons other than to mitigate the consequences of an event (e.g., at the discretion of the licensee as part of a planned procedure or evolution).

Sections 50.73(a)(2) (v) and (vi) (proposed § 50.73(a)(2)) require reporting of:

* * * *

(v) Any event or condition that alone could have prevented the fulfillment of the safety function of structures or systems that are needed to:

(A) Shut down the reactor and maintain it in a safe shutdown condition;

(B) Remove residual heat;

(C) Control the release of radioactive material; or

(D) Mitigate the consequences of an accident.

(vi) Events covered in paragraph (a)(2)(v) of this section may include one or more personnel errors, equipment failures, and/or discovery of design, analysis, fabrication, construction, and/or procedural inadequacies. However, individual component failures need not be reported pursuant to this paragraph if redundant equipment in the same system was operable and available to perform the required safety function.

The wording of this paragraph has been changed from the proposed rule to make it easier to read. The intent and scope of the paragraph have not been changed.

The intent of this paragraph is to capture those events where there would have been a failure of a safety system to properly complete a safety function, regardless of when the failures were discovered or whether the system was needed at the time.

This paragraph is also based on the assumption that safety-related systems and structures are intended to mitigate the consequences of an accident. While § 50.73(a)(2)(iv) of this final rule applies to actual actuations of an ESF, § 50.73(a)(2)(v) of this final rule covers an event or condition where redundant structures, components, or trains of a safety system could have failed to perform their intended function because of: one or more personnel errors, including procedure violations; equipment failures; or design, analysis, fabrication, construction, or procedural deficiencies. The event must be reported regardless of the situation or condition that caused the structure or systems to be unavailable, and regardless of whether or not an alternate safety system could have been used to perform the safety function (e.g., High Pressure Core Cooling failed, but feed-and-bleed or Low Pressure Core Cooling were available to provide the safety function of core cooling).

The applicability of this paragraph includes those safety systems designed to mitigate the consequences of an accident (e.g., containment isolation, emergency filtration). Hence, minor operational events involving a specific component such as valve packing leaks, which could be considerd a lack of control of radioactive material, should not be reported under this paragraph. System leaks or other similar events may, however, be reportable under other paragraphs.

It should be noted that there are a limited number of single-train systems that perform safety functions (e.g., the High Pressure Coolant Injection System in BWRs). For such systems, loss of the single train would prevent the fulfillment of the safety function of that system and, therefore, must be reported even though the plant Technical Specifications may allow such a condition to exist for a specified limited length of time.

It should also be noted that, if a potentially serious human error is made that could have prevented fulfillment of a safety function, but recovery factors resulted in the error being corrected, the error is still reportable.

The Commission recognizes that the application of this and other paragraphs of this section involves the use of engineering judgment on the part of licensees. In this case, a technical judgment must be made whether a failure or operator action that did actually disable one train of a safety system, could have, but did not, affect a redundant train within the ESF system. If so, this would constitute an event that "could have prevented" the fulfillment of a safety function, and, accordingly, must be reported.

If a component fails by an apparently random mechanism it may or may not be reportable if the functionally redundant component could fail by the same mechanism. Reporting is required if the failure constitutes a condition where there is reasonable doubt that the functionally redundant train or channel would remain operational until it completed its safety function or is repaired. For example, if a pump in one train of an ESF system fails because of improper lubrication, and engineering judgment indicates that there is a reasonable expectation that the functionally redundant pump in the other train, which was also improperly lubricated, would have also failed before it completed its safety function, then the actual failure is reportable and the potential failure of the functionally redundant pump must be discussed in the LER.

For safety systems that include three or more trains, the failure of two or more trains should be reported if, in the judgement of the licensee, the functional capability of the overall system was jeopardized.

Interaction between systems, particularly a safety system and a non-safety system, is also included in this criterion. For example, the Commission is increasingly concerned about the effect of a loss or degradation of what had been assumed to be non-essential inputs to safety systems. Therefore, this paragraph also includes those cases where a service (e.g., heating, ventilation, and cooling) or input (e.g., compressed air) which is necessary for reliable or long-term operation of a safety system is lost or degraded. Such loss or degradation is reportable if the proper fulfillment of the safety function is not cannot be assured. Failures that affect inputs or services to systems that have no safety function need not be reported.

Finally the Commission recognizes that the licensee may also use engineering judgment to decide when personnel actions *could* have prevented fulfillment of a safety function. For example, when an individual improperly operates or maintains a component, he might conceivably have made the same error for all of the functionally redundant components (e.g., if he incorrectly calibrates one bistable amplifier in the Reactor Protection System, he could conceivably incorrectly calibrate all bistable amplifiers). However, for an event to be reportable it is necessary that the actions actually affect or involve components in more than one train or channel of a safety system, and the result of the actions must be undesirable from the perspective of protecting the health and safety of the public. The components can be functionally redundant (e.g., two pumps in different trains) or not functionally redundant (e.g., the operator correctly stops a pump in Train "A" and, instead of shutting the pump discharge valve in Train "A," he mistakenly shuts the pump discharge valve in Train "B").

Section 50.73(a)(2)(vii) (proposed § 50.73(a)(3)) requires the reporting of: "Any event where a single cause or condition caused at least one independent train of channel to become inoperable in multiple systems or two independent trains channels or to become inoperable in a system designed to:

(A) Shut down the reactor and maintain it in a safe shutdown condition.

(B) Remove residual heat.

(C) Control the release of radioactive material; or

(D) Mitigate the consequences of an accident."

This paragraph has been changed to clarify the intent of the phrase, "nonconservative interdependence." Numerous comment letters expressed difficulty in understanding what this phrase meant, so the paragraph has been changed to be more specific. The new paragraph is narrower in scope than the original paragraph because the term is specifically defined, but the basic intent is the same.

This paragraph requires those events to be reported where a single cause produced a component or group of components to become inoperable in redundant or independent portions (i.e., trains or channels) of one or more systems having a safety function. These events can identify previously unrecognized common cause failures and systems interactions. Such failures can be simultaneous failures which occur because of a single initiating cause (i.e., the single cause or mechanism serves as a common input to the failures); or the failures can be sequential (i.e., cascade failures), such as the case where a single component failure results in the failure of one or more additional components.

To be reportable, however, the event or failure must result in or involve the failure of independent portions of more than one train or channel in the same or different systems. For example, if a cause or condition caused components in Train "A" and "B" of a single system to become inoperable, even if additional trains (e.g., Train "C") were still available, the event must be reported. In addition, if the cause or condition caused components in Train "A" of one system and in Train "B" of another system (i.e., a train that is assumed in the safety analysis to be independent) to become inoperable, the event must be reported. However, if a cause or condition caused components in Train "A" of one system and Train "A" of another system (i.e., trains that are not assumed in the safety analysis to be independent), the event need not be reported unless it meets one or more of the other criteria in this section.

In addition, this paragraph does not include those cases where one train of a system or a component was removed from service as part of a planned evolution, in accordance with an approved procedure, and in accordance with the plant's Technical Specifications. For example, if the licensee removes part of a system from service to perform maintenance, and the Technical Specifications permit the resulting configuration, and the system or component is returned to service

within the time limit specified in the Technical Specifications, the action need not be reported under this paragraph. However, if, while the train or component is out of service, the licensee identifies a condition that could have prevented the whole system from performing its intended function (e.g., the licensee finds a set of relays that is wired incorrectly), that condition must be reported.

Section 50.73(a)(2)(i) (proposed § 50.73(a)(4)) requires reporting of:

"(A) The completion of any nuclear plant shutdown required by the plant's Technical Specifications; or

"(B) Any operation prohibited by the plant's Technical Specifications; or

"(C) Any deviation from the plant's Technical Specifications authorized pursuant to § 50.54(x) of this part."

This paragraph has been reworded to more clearly define the events that must be reported. In addition, the scope has been changed to require the reporting of events or conditions "prohibited by the plant's Technical Specifications" rather than events where "a plant Technical Specification Action Statement is not met." This change accommodates plants that do not have requirements that are specifically defined as Action Statements.

This paragraph now requires events to be reported where the licensee is required to shut down the plant because the requirements of the Technical Specifications were not met. For the purpose of this paragraph, "shutdown" is defined as the point in time where the Technical Specifications require that the plant be in the first shutdown condition required by a Limiting Condition for Operation (e.g., hot standby (Mode 3) for PWRs with the Standard Technical Specifications). If the condition is corrected before the time limit for being shut down (i.e., before completion of the shutdown), the event need not be reported.

In addition, if a condition that was prohibited by the Technical Specifications existed for a period of time longer than that permitted by the Technical Specifications, it must be reported even if the condition was not discovered until after the allowable time had elapsed and the condition was rectified immediately after discovery.

Section 50.73(a)(2)(ii) (proposed § 50.73(a)(5)) requires reporting of: "Any event or condition that resulted in the condition of the nuclear power plant, including its principal safety barriers, being seriously degraded, or that resulted in the nuclear power plant being:

"(A) In an unanalyzed condition that significantly compromised plant safety;

"(B) In a condition that was outside the design basis of the plant; or

"(C) In a condition not covered by the plant's operating and emergency procedures."

This paragraph requires events to be reported where the plant, including its principal safety barriers, was seriously degraded or in an unanalyzed condition.

For example, small voids in systems designed to remove heat from the reactor core which have been previously shown through analysis not to be safety significant need not be reported. However, the accumulation of voids that could inhibit the ability to adequately remove heat from the reactor core, particularly under natural circulation conditions, would constitute an unanalyzed condition and must be reported. In addition, voiding in instrument lines that results in an erroneous indication causing the operator to significantly misunderstand the true condition of the plant is also an unanalyzed condition and must be reported.

The Commission recognizes that the licensee may use engineering judgment and experience to determine whether an unanalyzed condition existed. It is not intended that this paragraph apply to minor variations in individual parameters, or to problems concerning single pieces of equipment. For example, at any time, one or more safety-related components may be out of service due to testing, maintenance, or a fault that has not yet been repaired. Any trivial single failure or minor error in performing surveillance tests could produce a situation in which two or more often unrelated, safety-related components are out-of-service. Technically, this is an unanalyzed condition. However, these events should be reported only if they involve functionally related components or if they significantly compromise plant safety.

Finally, this paragraph also includes material (e.g., metallurgical, chemical) problems that cause abnormal degradation of the principal safety barriers (i.e., the fuel cladding, reactor coolant system pressure boundary, or the containment).

Additional examples of situations included in this paragraph are:

(a) Fuel cladding failures in the reactor or in the storage pool, that exceed expected values, that are unique or widespread, or that resulted from unexpected factors.

(b) Reactor coolant radioactivity levels that exceeded Technical Specification limits for iodine spikes or,

radioactivity levels at a BWR air ejector monitor that exceeded the Technical Specification limits.

(c) Cracks and breaks in piping, the reactor vessel, or major components in the primary coolant circuit that have safety relevance (steam generators, reactor coolant pumps, valves, etc.).

(d) Significant welding or material defects in the primary coolant system.

(e) Serious temperature or pressure transients (e.g., transients that violate the plant's Technical Specifications).

(f) Loss of relief and/or safety valve operability during test or operation (such that the number of operable valves or man-way closures is less than required by the Technical Specifications).

(g) Loss of containment function or integrity (e.g., containment leakage rates exceeding the authorized limits).

Section 50.73(a)(2)(iii) (proposed § 50.73(a)(6)) requires reporting of: "Any natural phenomenon or other external condition that posed an actual threat to the safety of the nuclear power plant or significantly hampered site personnel in the performance of duties necessary for the safe operation of the nuclear power plant."

This paragraph has been reworded to make it clear that it applies only to acts of nature (e.g., tornadoes) and external hazards (e.g., railroad tank car explosion). References to acts of sabotage have been removed because they are covered by § 73.71. In addition, threats to personnel from internal hazards (e.g., radioactivity releases) are now covered by a separate paragraph (§ 50.73(a)(2)(x)).

This paragraph requires those events to be reported where there is an actual threat to the plant from an external condition or natural phenomenon, and where the threat or damage challenges the ability of the plant to continue to operate in a safe manner (including the orderly shutdown and maintenance of shutdown conditions).

The licensee is to decide if a phenomenon or condition actually threatened the plant. For example, a minor brush fire in a remote area of the site that was quickly controlled by fire fighting personnel and, as a result, did not present a threat to the plant need not be reported. However, a major forest fire, large-scale flood, or major earthquake that presents a clear threat to the plant must be reported. Industrial or transportation accidents that occurred near the site and created a plant safety concern must also be reported.

Section 50.73(a)(2)(x) (proposed § 50.73(a)(7)) requires reporting of: "Any event that posed an actual threat to the

safety of the nuclear power plant or significantly hampered site personnel in the performance of duties necessary for the safe operation of the nuclear power plant including fires, toxic gas releases, or radioactive releases."

This paragraph has been reworded to include physical hazards (internal to the plant) to personnel (e.g., electrical fires). In addition, in response to numerous comments, the scope has been narrowed so that the hazard must hamper the ability of site personnel to perform safety-related activities affecting plant safety.

In-plant releases must be reported if they require evacuation of rooms or buildings containing systems important to safety and, as a result, the ability of the operators to perform necessary safety functions is significantly hampered. Precautionary evacuations of rooms and buildings that subsequent evaluation determines were not required need not be reported.

Proposed § 50.73(a)(8) was intended to capture an event that involved a controlled release of a significant amount of radioactive material to offsite areas. In addition, "significant" was based on the plant's Technical Specification limits for the release of radioactive material. However, this section has been deleted because the reporting of these events is already required by § 50.73(a)(2)(i) and § 20.405.

Section 50.73(a)(2) (viii) and (ix) (proposed § 50.73(a)(9)) require reporting of:

*	*	*	*	*

(viii)(A) Any airborne radioactivity release that exceeded 2 times the applicable concentrations of the limits specified in Table II of Appendix B to Part 20 of this chapter in unrestricted areas, when averaged over a time period of one hour.

(B) Any liquid effluent release that exceeded 2 times the limiting combined Maximum Permissible Concentration (MPC) (see Note 1 of Appendix B to Part 20 of this chapter) at the point of entry into the receiving water (i.e., unrestricted area) for all radionuclides except tritium and dissolved noble gases, when averaged over a time period of one hour.

(ix) Reports submitted to the Commission in accordance with paragraph (a)(2)(viii) of this section also meet the effluent release reporting requirements of paragraph 20.405(a)(5) of Part 20 of this chapter.

*	*	*	*	*

Paragraph (viii) has been changed to clarify the requirements to report releases of radioactive material. The paragraph is similar to § 20.405 but places a lower threshold for reporting events at commercial power reactors. The lower threshold is based on the significance of the breakdown of the licensee's program necessary to have a release of this size, rather than on the significance of the impact of the actual release.

Reports of events covered by § 50.73(a)(2)(viii) are to be made in lieu of reporting noble gas releases that exceed 10 times the instantaneous release rate, without averaging over a time period, as implied by the requirement of § 20.405(a)(5).

Paragraph 50.73(b) describes the format and content of the LER. It requires that the licensee prepare the LER in sufficient depth so that knowledgeable readers conversant with the design of commercial nuclear power plants, but not familiar with the details of a particular plant, can understand the complete event (i.e., the cause of the event, the plant status before the event, and the sequence of occurrences during the event).

Paragraph 50.73(b)(1) requires that the licensee provide a brief abstract describing the major occurrences during the event, including all actual component or system failures that contributed to the event, all relevant operator errors or violations of procedures, and any significant corrective action taken or planned as a result of the event. This paragraph is needed to give LER data base users a brief description of the event in order to identify events of interest.

Paragraph 50.73(b)(2) requires that the licensee include in the LER a clear, specific narrative statement of exactly what happened during the entire event so that readers not familiar with the details of a particular plant can understand the event. The licensee should emphasize how systems, components, and operating personnel performed. Specific hardware problems should not be covered in excessive detail. Characteristics of a plant that are unique and that influenced the event (favorably or unfavorably) must be described. The narrative must also describe the event from the perspective of the operator (e.g., what the operator saw, did, perceived, understood, or misunderstood).

Paragraph 50.73(b)(3) requires that the LER include a summary assessment of the actual and potential safety consequences and implications of the event. This assessment may be based on the conditions existing at the time of the event. The evaluation must be carried out to the extent necessary to fully assess the safety consequences and safety margins associated with the event. An assessment of the event under alternative conditions must be included if the incident would have been more severe (e.g., the plant would have been

in a condition not analyzed in the Safety Analysis Report) under reasonable and credible alternative conditions, such as power level or operating mode. For example, if an event occurred while the plant was at 15% power and the same event could have occurred while the plant was at 100% power, and, as a result, the consequences would have been considerably more serious, the licensee must assess and report those consequences.

Paragraph 50.73(b)(4) requires that the licensee describe in the LER any corrective actions planned as a result of the event that are known at the time the LER is submitted, including actions to reduce the probability of similar events occurring in the future. After the initial LER is submitted only substantial changes in the corrective action need be reported as a supplemental LER.

Paragraph 50.73(c) authorizes the NRC staff to require the licensee to submit specific supplemental information beyond that required by § 50.73(b). Such information may be required if the staff finds that supplemental material is necessary for complete understanding of an unusually complex or significant event. Such requests for supplemental information must be made in writing, and the licensee must submit the requested information as a supplement to the initial LER within a time period mutually agreed upon by the NRC staff and the licensee.

Paragraph 50.73(f) gives the NRC's Executive Director for Operations the authority to grant case-by-case exemptions to the reporting requirements contained in the LER system. This exemption could be used to limit the collection of certain data in those cases where full participation would be unduly difficult because of a plant's unique design or circumstances.

Paragraph 50.73(g) states that the reporting requirements contained in § 50.73 replace the reporting requirements in all nuclear power plant Technical Specifications that are typically associated with Reportable Occurrences.

The reporting requirements superseded by § 50.73 are those contained in the Technical Specification sections that are usually titled "Prompt Notification with Written Followup" (Section 6.9.1.8) and "Thirty Day Written Reports" (Section 6.9.1.9). The reporting requirements that have been superseded are also described in Regulatory Guide 1.16, Revision 4, "Reporting of Operating Information—Appendix A Technical Specification," Paragraph 2, "Reportable Occurrences." The special report typically described in Section 6.9.2

"Special Reports" of the Technical Specifications are still required.

V. Regulatory Analysis

The Commission has prepared a regulatory analysis for this final rule. The analysis examines the costs and benefits of the alternatives considered by the Commission. A copy of the regulatory analysis is available for inspection and copying for a fee at the NRC Public Document Room, 1717 H Street, N.W., Washington, D.C. Single copies of the analysis may be obtained from Frederick J. Hebon, Chief, Program Technology Branch, Office for Analysis and Evaluation of Operational Data, U.S. Nuclear Regulatory Commission, Washington, D.C. 20555; Telephone (301) 492–4480.

VI. Paperwork Reduction Act Statement

The Nuclear Regulatory Commission has submitted this rule to the Office of Management and Budget for such review as may be appropriate under the Paperwork Reduction Act, Pub. L. 96–511. The date on which the reporting requirements of this rule become effective reflects inclusion of the 60-day period which the Act allows for such review.

VII. Regulatory Flexibility Certification

In accordance with the Regulatory Flexibility Act of 1980, 5 U.S.C. 605(b), the Commission hereby certifies that this rule will not have a significant economic impact on a substantial number of small entities. This final rule affects electric utilities that are dominant in their respective service areas and that own and operate nuclear utilization facilities licensed under sections 103 and 104b of the Atomic Energy Act of 1954, as amended. The amendments clarify and modify presently existing notification requirements.

Accordingly, there is no new, significant economic impact on these licensees, nor do these licensees fall within the scope of the definition of "small entities" set forth in the Regulatory Flexibility Act or the Small Business Size Standards set out in regulations issued by the Small Business Administration at 13 CFR Part 121.

List of Subjects

10 CFR Part 20

Licensed material, Nuclear power plants and reactors, Penalty, Reporting and recordkeeping requirements.

10 CFR PARTS 50

Incorporation by reference, Antitrust, Classified information, Fire protection.

Intergovernmental relations, Nuclear power plants and reactors, Penalty, Radiation protection, Reporting and recordkeeping requirements. --

Under the authority of the Atomic Energy Act of 1954, as amended, the Energy Reorganization Act of 1974, as amended, and 5 U.S.C. 552 and 553, the following amendments to 10 CFR Parts 20 and 50 are published as a document subject to codification.

PART 50—DOMESTIC LICENSING OF PRODUCTION AND UTILIZATION FACILITIES

1. The authority citation for Part 50 continues to read as follows:

Authority: Secs. 103, 104, 161, 182, 183, 186, 189, 68 Stat. 936, 937, 948, 953, 954, 955, 958, as amended, sec. 234, 83 Stat. 1244, as amended (42 U.S.C. 2133, 2134, 2201, 2232, 2233, 2236, 2239, 2282); secs. 201, 202, 206, 68 Stat. 1242, 1244, 1246, as amended (42 U.S.C. 5841, 5842, 5846), unless otherwise noted.

Section 50.7 also issued under Pub. L. 95-601, sec. 10, 92 Stat. 2951 (42 U.S.C. 5851). Sections 50.58, 50.91 and 50.92 also issued under Pub. L. 97-415, 96 Stat. 2073 (42 U.S.C. 2239). Section 50.78 also issued under sec. 122, 68 Stat. 939 (42 U.S.C. 2152). Sections 50.80-50.81 also issued under sec. 184, 68 Stat. 954, as amended (42 U.S.C. 2234). Sections 50.100-50-102 also issued under sec. 186, 68 Stat. 955 (42 U.S.C. 2236).

For the purposes of sec. 223, 68 Stat. 958, as amended (42 U.S.C. 2273), §§ 50.10 (a), (b), and (c), 50.44, 50.46, 50.48, 50.54, and 50.80(a) are issued under sec. 161b, 68 Stat. 948, as amended (42 U.S.C. 2201(b)); §§ 50.10 (b) and (c) and 50.54 are issued under sec. 161i, 68 Stat. 949, as amended (42 U.S.C. 2201(i)); and §§ 50.55(e), 50.59(b), 50.70, 50.71, 50.72, and 50.78 are issued under sec. 161o, 68 Stat. 950, as amended (42 U.S.C. 2201(o)).

2. A new § 50.73 is added to read as follows:

§ 50.73   Licensee event report system.

(a) *Reportable events.* (1) The holder of an operating license for a nuclear power plant (licensee) shall submit a Licensee Event Report (LER) for any event of the type described in this paragraph within 30 days after the discovery of the event. Unless otherwise specified in this section, the licensee shall report an event regardless of the plant mode or power level, and regardless of the significance of the structure, system, or component that initiated the event.

(2) The licensee shall report:

(i)(A) The completion of any nuclear plant shutdown required by the plant's Technical Specifications; or

(B) Any operation or condition prohibited by the plant's Technical Specifications; or

(C) Any deviation from the plant's Technical Specifications authorized pursuant to § 50.54(x) of this part.

(ii) Any event or condition that resulted in the condition of the nuclear power plant, including its principal safety barriers, being seriously degraded, or that resulted in the nuclear power plant being:

(A) In an unanalyzed condition that significantly compromised plant safety;

(B) In a condition that was outside the design basis of the plant; or

(C) In a condition not covered by the plant's operating and emergency procedures.

(iii) Any natural phenomenon or other external condition that posed an actual threat to the safety of the nuclear power plant or significantly hampered site personnel in the performance of duties necessary for the safe operation of the nuclear power plant.

(iv) Any event or condition that resulted in manual or automatic actuation of any Engineered Safety Feature (ESF), including the Reactor Protection System (RPS). However, actuation of an ESF, including the RPS, that resulted from and was part of the preplanned sequence during testing or reactor operation need not be reported.

(v) Any event or condition that alone could have prevented the fulfillment of the safety function of structures or systems that are needed to:

(A) Shut down the reactor and maintain it in a safe shutdown condition;

(B) Remove residual heat;

(C) Control the release of radioactive material; or

(D) Mitigate the consequences of an accident.

(vi) Events covered in paragraph (a)(2)(v) of this section may include one or more procedural errors, equipment failures, and/or discovery of design, analysis, fabrication, construction, and/or procedural inadequacies. However, individual component failures need not be reported pursuant to this paragraph if redundant equipment in the same system was operable and available to perform the required safety function.

(vii) Any event where a single cause or condition caused at least one independent train or channel to become inoperable in multiple systems or two independent trains or channels to become inoperable in a single system designed to:

(A) Shut down the reactor and maintain it in a safe shutdown condition;

(B) Remove residual heat;

(C) Control the release of radioactive material; or

(D) Mitigate the consequences of an accident.

(viii)(A) Any airborne radioactivity release that exceeded 2 times the applicable concentrations of the limits specified in Appendix B, Table II of Part 20 of this chapter in unrestricted areas, when averaged over a time period of one hour.

(B) Any liquid effluent release that exceeded 2 times the limiting combined Maximum Permissible Concentration (MPC) (see Note 1 of Appendix B to Part 20 of this chapter) at the point of entry into the receiving water (i.e., unrestricted area) for all radionuclides except tritium and dissolved noble gases, when averaged over a time period of one hour.

(ix) Reports submitted to the Commission in accordance with paragraph (a)(2)(viii) of this section also meet the effluent release reporting requirements of paragraph 20.405(a)(5) of Part 20 of this chapter.

(x) Any event that posed an actual threat to the safety of the nuclear power plant or significantly hampered site personnel in the performance of duties necessary for the safe operation of the nuclear power plant including fires, toxic gas releases, or radioactive releases.

(b) *Contents.* The Licensee Event Report shall contain:

(1) A brief abstract describing the major occurrences during the event, including all component or system failures that contributed to the event and significant corrective action taken or planned to prevent recurrence.

(2)(i) A clear, specific, narrative description of what occurred so that knowledgeable readers conversant with the design of commercial nuclear power plants, but not familiar with the details of a particular plant, can understand the complete event.

(ii) The narrative description must include the following specific information as appropriate for the particular event:

(A) Plant operating conditions before the event.

(B) Status of structures, components, or systems that were inoperable at the start of the event and that contributed to the event.

(C) Dates and approximate times of occurrences.

(D) The cause of each component or system failure or personnel error, if known.

(E) The failure mode, mechanism, and effect of each failed component, if known.

(F) The Energy Industry Identification System component function identifier and system name of each component or system referred to in the LER.

(1) The Energy Industry Identification System is defined in: IEEE Std 803–1983 (May 16, 1983) Recommended Practices for Unique Identification Plants and Related Facilities—Principles and Definitions.

(2) IEEE Std 803–1983 has been approved for incorporation by reference by the Director of the Federal Register. A notice of any changes made to the material incorporated by reference will be published in the Federal Register. Copies may be obtained from the Institute of Electrical and Electronics Engineers, 345 East 47th Street, New York, NY 10017. A copy is available for inspection and copying for a fee at the Commission's Public Document Room, 1717 H Street, NW., Washington, D.C. and at the Office of the Federal Register, 1100 L St. NW., Washington, D.C.

(G) For failures of components with multiple functions, include a list of systems or secondary functions that were also affected.

(H) For failure that rendered a train of a safety system inoperable, an estimate of the elapsed time from the discovery of the failure until the train was returned to service.

(I) The method of discovery of each component or system failure or procedural error.

(J)(1) Operator actions that affected the course of the event, including operator errors, procedural deficiencies, or both, that contributed to the event.

(2) For each personnel error, the licensee shall discuss:

(i) Whether the error was a cognitive error (e.g., failure to recognize the actual plant condition, failure to realize which systems should be functioning, failure to recognize the true nature of the event) or a procedural error;

(ii) Whether the error was contrary to an approved procedure, was a direct result of an error in an approved procedure, or was associated with an activity or task that was not covered by an approved procedure;

(iii) Any unusual characteristics of the work location (e.g., heat, noise) that directly contributed to the error; and

(iv) The type of personnel involved (i.e., contractor personnel, utility-licensed operator, utility nonlicensed operator, other utility personnel).

(K) Automatically and manually initiated safety system responses.

(L) The manufacturer and model number (or other identification) of each component that failed during the event.

(3) An assessment of the safety consequences and implications of the event. This assessment must include the availability of other systems or components that could have performed the same function as the components and systems that failed during the event.

(4) A description of any corrective actions planned as a result of the event, including those to reduce the probability of similar events occurring in the future.

(5) Reference to any previous similar events at the same plant that are known to the licensee.

(6) The name and telephone number of a person within the licensee's organization who is knowledgeable about the event and can provide additional information concerning the event and the plant's characteristics.

(c) Supplemental information. The Commission may require the licensee to submit specific additional information beyond that required by paragraph (b) of this section if the Commission finds that supplemental material is necessary for complete understanding of an unusually complex or significant event. These requests for supplemental information will be made in writing and the licensee shall submit the requested information as a supplement to the initial LER.

(d) Submission of reports. Licensee Event Reports must be prepared on Form NRC 366 and submitted within 30 days of discovery of a reportable event or situation to the U.S. Nuclear Regulatory Commission, Document Control Desk, Washington, D.C. 20555. The licensee shall also submit an additional copy to the appropriate NRC Regional Office listed in Appendix A to Part 73 of this chapter.

(e) Report legibility. The reports and copies that licensees are required to submit to the Commission under the provisions of this section must be of sufficient quality to permit legible reproduction and micrographic processing.

(f) Exemptions. Upon written request from a licensee including adequate justification or at the initiation of the NRC staff, the NRC Executive Director for Operations may, by a letter to the licensee, grant exemptions to the reporting requirements under this section.

(g) Reportable occurrences. The requirements contained in this section replace all existing requirements for licensees to report "Reportable Occurrences" as defined in individual plant Technical Specifications.

The following additional amendments are also made to Parts 20 and 50 of the regulations in this chapter.

PART 20—STANDARDS FOR PROTECTION AGAINST RADIATION

3. In § 20.402, paragraph (a) is revised; the introductory text of paragraph (b) is revised; and a new paragraph (e) is added to read as follows:

§ 20.402 Reports of theft or loss of licensed material.

(a)(1) Each licensee shall report to the Commission, by telephone, immediately after it determines that a loss or theft of licensed material has occurred in such quantities and under such circumstances that it appears to the licensee that a substantial hazard may result to persons in unrestricted areas.

(2) Reports must be made as follows:

(i) Licensees having an installed Emergency Notification System shall make the reports to the NRC Operations Center in accordance with § 50.72 of this chapter.

(ii) All other licensees shall make reports to the Administrator of the appropriate NRC Regional Office listed in Appendix D of this part.

(b) Each licensee who makes a report under paragraph (a) of this section shall, within 30 days after learning of the loss or theft, make a report in writing to the U.S. Nuclear Regulatory Commission, Document Control Desk, Washington, D.C. 20555, with a copy to the appropriate NRC Regional Office listed in Appendix D of this part. The report shall include the following information:

*   *   *   *   *

(e) For holders of an operating license for a nuclear power plant, the events included in paragraph (b) of this section must be reported in accordance with the procedures described in § 50.73 (b), (c), (d), (e), and (g) of this chapter and must include the information required in paragraph (b) of this section. Events reported in accordance with § 50.73 of this chapter need not be reported by a duplicate report under paragraph (b) of this section.

4. In § 20.403, the introductory text of paragraphs (a) and (b) is revised, and paragraph (d) is revised to read as follows:

§ 20.403 Notifications of incidents.

(a) Immediate notification. Each licensee shall immediately report any events involving byproduct, source, or special nuclear material possessed by the licensee that may have caused or threatens to cause:

*   *   *   *   *

(b) Twenty-four hour notification. Each licensee shall within 24 hours of discovery of the event, report any event involving licensed material possessed

by the licensee that may have caused or threatens to cause:

* * * * *

(d) Reports made by licensees in response to the requirements of this section must be made as follows:

(1) Licensees that have an installed Emergency Notification System shall make the reports required by paragraphs (a) and (b) of this section to the NRC Operations Center in accordance with § 50.72 of this chapter.

(2) All other licensees shall make the reports required by paragraphs (a) and (b) of this section by telephone and by telegram, mailgram, or facsimile to the Administrator of the appropriate NRC Regional Office listed in Appendix D of this part.

5. In § 20.405, paragraphs (a) and (c) are revised, and new paragraphs (d) and (e) are added to read as follows:

§ 20.405   Reports of overexposures and excessive levels and concentrations.

(a)(1) In addition to any notification required by § 20.403 of this part, each licensee shall make a report in writing concerning any one of the following types of incidents within 30 days of its occurrence:

(i) Each exposure of an individual to radiation in excess of the applicable limits in §§ 20.101 or 20.104(a) of this part, or the license;

(ii) Each exposure of an individual to radioactive material in excess of the applicable limits in §§ 20.103(a)(1), 20.103(a)(2), or 20.104(b) of this part, or in the license;

(iii) Levels of radiation or concentrations of radioactive material in a restricted area in excess of any other applicable limit in the license;

(iv) Any incident for which notification is required by § 20.403 of this part; or

(v) Levels of radiation or concentrations of radioactive material (whether or not involving excessive exposure of any individual) in an unrestricted area in excess of ten times any applicable limit set forth in this part or in the license.

(2) Each report required under paragraph (a)(1) of this section must describe the extent of exposure of individuals to radiation or to radioactive material, including:

(i) Estimates of each individual's exposure as required by paragraph (b) of this section:

(ii) Levels of radiation and concentrations of radioactive material involved;

(iii) The cause of the exposure, levels or concentrations; and

(iv) Corrective steps taken or planned to prevent a recurrence.

* * * * *

(c)(1) In addition to any notification required by § 20.403 of this part, each licensee shall make a report in writing of levels of radiation or releases of radioactive material in excess of limits specified by 40 CFR Part 190, "Environmental Radiation Protection Standards for Nuclear Power Operations," or in excess of license conditions related to compliance with 40 CFR Part 190.

(2) Each report submitted under paragraph (c)(1) of this section must describe:

(i) The extent of exposure of individuals to radiation or to radioactive material;

(ii) Levels of radiation and concentrations of radioactive material involved;

(iii) The cause of the exposure, levels, or concentrations; and

(iv) Corrective steps taken or planned to assure against a recurrence, including the schedule for achieving conformance with 40 CFR Part 190 and with associated license conditions.

(d) For holders of an operating license for a nuclear power plant, the incidents included in paragraphs (a) or (c) of this section must be reported in accordance with the procedures described in paragraphs 50.73 (b), (c), (d), (e), and (g) of this chapter and must also include the information required by paragraphs (a) and (c) of this section. Incidents reported in accordance with § 50.73 of this chapter need not be reported by a duplicate report under paragraphs (a) or (c) of this section.

(e) All other licensees who make reports under paragraphs (a) or (c) of this section shall, within 30 days after learning of the overexposure or excessive level or concentration, make a report in writing to the U.S. Nuclear Regulatory Commission, Document Control Desk, Washington, D.C. 20555, with a copy to the appropriate NRC Regional Office listed in Appendix D of this part.

PART 50—DOMESTIC LICENSING OF PRODUCTION AND UTILIZATION FACILITIES

6. In § 50.36, new paragraphs (c)(6) and (7) are added to read as follows:

§ 50.36   Technical specifications.

* * * * *

(c) * * *

(6) *Initial Notification.* Reports made to the Commission by licensees in response to the requirements of this section must be made as follows:

(i) Licensees that have an installed Emergency Notification System shall make the initial notification to the NRC Operations Center in accordance with § 50.72 of this part.

(ii) All other licensees shall make the initial notification by telephone to the Administrator of the appropriate NRC Regional Office listed in Appendix D, Part 20, of this chapter.

(7) *Written reports.* Holders of an operating license for a nuclear power-plant shall submit a written report to the Commission concerning the incidents included in paragraphs (c) (1) and (2) of this section in accordance with the procedures described in § 50.73 (b), (c), (d), (e), and (g) of this part. Incidents reported in accordance with § 50.73 of this part need not also be reported under paragraphs (c) (1) or (2) of this section.

Dated at Washington, D.C. this 20th day of July 1983.

For the Nuclear Regulatory Commission.

Samuel J. Chilk,

Secretary of the Commission.

[FR Doc. 83-20166 Filed 7-25-83; 8:45 am]

BILLING CODE 7590-01-M

APPENDIX F

1992 REVISION TO 10 CFR 50.72 AND 50.73 INCLUDING
STATEMENT OF CONSIDERATIONS

Published in the *Federal Register*
on September 9, 1992 September 10, 1992
(Vol. 57, No. 176, pages 41378-41381)

NOTE: This *Federal Register* notice does not provide a complete version of 10 CFR
50.72 and 50.73; it addresses only small parts of those sections. Its purpose
here is to present the Statement of Considerations, which explains some of the
reporting requirements of the sections.

certified to OMB, in a letter dated August 14, 1992, that by unanimous vote the Commission had overridden the OMB's disapproval of the information collection request associated with this rule.

On August 21, 1992, OMB assigned the following new control number: 3150–0171, effective until August 31, 1995.

This new control number is only applicable to the sections in 10 CFR part 35 amended by this rule. Information collection authority for all other sections of 10 CFR part 35 remains under the existing general control number: 3150–0010.

List of Subjects in 10 CFR Part 35

Byproduct material, Criminal penalty, Drugs, Health facilities, Health professions, Incorporation by reference, Medical devices, Nuclear materials, Occupational safety and health, Radiation protection, Reporting and recordkeeping requirements.

Text of Final Regulations

For the reasons set out in the preamble and under the authority of the Atomic Energy Act of 1954, as amended, the Energy Reorganization Act of 1974, as amended, and 5 U.S.C. 552 and 553, the NRC is adopting the following amendments to 10 CFR part 35.

PART 35—MEDICAL USE OF BYPRODUCT MATERIAL

1. The authority citation for part 35 continues to read in part as follows:

Authority: Secs. 161, 68 Stat. 948, as amended (42 U.S.C. 2201); sec. 201, 88 Stat. 1242, as amended (42 U.S.C. 5841) * * *.

2. In § 35.8, paragraph (b) is revised and paragraph (d) is added to read as follows:

§ 35.8 Information collection requirements: OMB approval.

*   *   *   *   *

(b) The approved information collection requirements contained in this part appear in §§ 35.12, 35.13, 35.14, 35.21, 35.22, 35.23, 35.27, 35.29, 35.31, 35.50, 35.51, 35.53, 35.59, 35.60, 35.61, 35.70, 35.80, 35.92, 35.204, 35.205, 35.310, 35.315, 35.404, 35.406, 35.410, 35.415, 35.606, 35.610, 35.615, 35.630, 35.632, 35.634, 35.636, 35.641, 35.643, 35.645, and 35.647.

*   *   *   *   *

(d) OMB has assigned control number 3150–0171 for the information collection requirements contained in §§ 35.32 and 35.33.

Dated at Rockville, Maryland, this 3d day of September 1992.

For the Nuclear Regulatory Commission.
Samuel J. Chilk,
Secretary of the Commission.
[FR Doc. 92–21754 Filed 9–9–92; 8:45 am]
BILLING CODE 7590–01–M

10 CFR Part 50

RIN 3150–AE12

Minor Modifications to Nuclear Power Reactor Event Reporting Requirements

AGENCY: Nuclear Regulatory Commission.

ACTION: Final rule.

SUMMARY: The Nuclear Regulatory Commission (NRC) has amended its regulations to make minor modifications to the current nuclear power reactor event reporting requirements. The final rule applies to all nuclear power reactor licensees and deletes reporting requirements for some events that have been determined to be of little or no safety significance. The final rule reduces the industry's reporting burden and the NRC's response burden in event review and assessment.

EFFECTIVE DATE: October 13, 1992.

FOR FURTHER INFORMATION CONTACT: Raji Tripathi, Office for Analysis and Evaluation of Operational Data, U.S. Nuclear Regulatory Commission, Washington, DC 20555. Telephone (301) 492–4435.

SUPPLEMENTARY INFORMATION:

Background

The Commission is issuing a final rule that amends the nuclear power reactor event reporting requirements contained in 10 CFR 50.72, "Immediate Notification Requirements for Operating Nuclear Power Reactors," and 10 CFR 50.73, "Licensee Event Report System." The final rule is issued as part of the Commission's ongoing activities to improve its regulations. Specifically, this final rule amends 10 CFR 50.72 (b)(2)(ii) and 10 CFR 50.73 (a)(2)(iv). On June 26, 1992 (57 FR 28642), the Commission issued a proposed rule requesting public comments on these amendments.

Over the past several years, the NRC has increased its attention to event reporting issues to ensure uniformity, consistency, and completeness in reporting. In September 1991, the NRC's Office for Analysis and Evaluation of Operational Data (AEOD) issued for comment a draft NUREG–1022, Revision 1,[1] "Event Reporting Systems 10 CFR

[1] Free single copy may be requested by writing to the Distribution and Mail Services Section, U.S.

50.72 and 10 CFR 50.73—Clarification of NRC Systems and Guidelines For Reporting." Following resolution of public comments, the NUREG will be issued in the final form. The NUREG will contain improved guidance for event reporting.

NRC's reviews of operating experience and the patterns of licensees' reporting of operating events since 1984 have indicated that reports on some of these events are not necessary for the NRC to perform its safety mission and that continued reporting of these events would not contribute useful information to the operating reactor events database. Additionally, these unnecessary reports would have continued to consume both the licensees' and the NRC's resources that could be better applied elsewhere. The NRC has determined that certain types of events, primarily those involving invalid engineered safety feature (ESF) actuations, are of little or no safety significance.

Valid ESF actuations are those actuations that result from "valid signals" or from intentional manual initiation, unless it is part of preplanned test. Valid signals are those signals that are initiated in response to actual plant conditions or parameters satisfying the requirements for ESF initiation.

Invalid actuations are by definition those that do not meet the criteria for being valid. Thus, invalid actuations include actuations that are not the result of valid signals and are not intentional manual actuations. Invalid actuations include instances where instrument drift, spurious signals, human error, or other invalid signals caused actuation of the ESF (e.g., jarring a cabinet, an error in use of jumpers or lifted leads, an error in actuation of switches or controls, equipment failure, or radio frequency interference).

NRC's evaluation of both the reported events since January 1984, when the existing rules first became effective, and the comments received during the Event Reporting Workshops conducted in Fall of 1990 identified needed improvements in the rules. The NRC determined that invalid actuation, isolation, or realignment of a limited set of ESFs including the systems, subsystems, or components (i.e., an invalid actuation, isolation, or realignment of only the reactor water clean-up (RWCU) system,

Nuclear Regulatory Commission, Washington, DC 20555. A copy is also available for inspection or copying for a fee at the NRC Public Document Room, 2120 L Street, NW, (Lower Level), Washington, DC 20555.

the control room emergency ventilation (CREV) system, the reactor building ventilation system, the fuel building ventilation system, or the auxiliary building ventilation system, or their equivalent ventilation systems) are of little or no safety significance. However, these events are currently reportable under 10 CFR 50.72 (b)(2)(ii) and 10 CFR 50.73 (a)(2)(iv).

The final rules for the current event reporting regulations, 10 CFR 50.72 and 10 CFR 50.73 (48 FR 39039; August 29, 1983, and 48 FR 33850; July 26, 1983, respectively), stated that ESF systems, including the reactor protection system (RPS), are provided to mitigate the consequences of a significant event. Therefore, ESFs should (1) work properly when called upon and (2) should not be challenged frequently or unnecessarily. The Statements of Consideration for these final rules also stated that operation of an ESF as part of a pre-planned operational procedure or test need not be reported. The Commission noted that ESF actuations, including reactor trips, are frequently associated with significant plant transients and are indicative of events that are of safety significance. At that time, the Commission also required all ESF actuations, including the RPS actuations, whether manual or automatic, valid or invalid—except as noted, to be reported to the NRC by telephone within 4 hours of occurrence followed by a written Licensee Event Report (LER) within 30 days of the incident. This requirement on timeliness of reporting remains unchanged.

The reported information is used by the NRC in confirmation of the licensing bases, identification of precursors to severe core damage, identification of plant specific deficiencies, generic lessons, review of management control systems, and licensee performance assessment.

Discussion

The NRC has determined that some events that involve only invalid ESF actuations are of little or no safety significance. However, not all invalid ESF actuations are being exempted from reporting through this rule. The relaxations in event reporting requirements contained in the final rule apply only to a narrow, limited set of specifically defined invalid ESF actuations. These events include invalid actuation, isolation, or realignment of a limited set of ESFs including systems subsystems, or components (i.e., an invalid actuation, isolation, or realignment of only the RWCU system, or the CREV system, reactor building ventilation system, fuel building

ventilation system, auxiliary building ventilation system, or their equivalent ventilation systems). The actuation of the standby gas treatment system following an invalid actuation of the reactor building ventilation system is also exempted from reporting. In addition, the final rule excludes invalid actuations of these ESFs (or their equivalent systems) from signals that originated from non-ESF circuitry.

However, invalid actuations of other ESFs would continue to be reportable. For example, emergency core cooling system isolations/actuations; containment isolation valve closures that affect cooling systems, main steam flow, essential support systems, etc., containment spray actuation; and residual heat removal system isolations (or systems designated by any other names but designed to fulfill the function similar to these systems and their equivalents), are still reportable. If an invalid ESF actuation reveals a defect in the system so that the system failed or would fail to perform its intended function, the event continues to be reportable under other requirements of 10 CFR 50.72 and 10 CFR 50.73. If a condition or deficiency has (1) an adverse impact on safety-related equipment and consequently on the ability to shut down the reactor and maintain it in a safe shutdown condition, (2) has a potential for significant radiological release or potential exposure to plant personnel or the general public, or (3) would compromise control room habitability, the event/discovery continues to be reportable.

Invalid ESF actuations that are excluded by this final rule, but occur as a part of a reportable event, continue to be described as part of the reportable event. These amendments are not intended to preclude submittal of a complete, accurate, and thorough description of an event that is otherwise reportable under 10 CFR 50.72 or 10 CFR 50.73. The Commission relaxed only the selected event reporting requirements specified in this final rule.

Licensees are still required under 10 CFR part 50, appendix B, "Quality Assurance Criteria for Nuclear Power Plants and Fuel Reprocessing Plants," to address corrective actions for events or conditions that are adverse to quality whether the event is reportable or not. In addition, minimizing ESF actuations (such as RWCU isolations) to reduce operational radiation exposures associated with the investigation and recovery from the actuations, are consistent with ALARA requirements.

This rule excludes three categories of events from reporting.

(1) The first category excludes events in which an invalid ESF or RPS actuation occurs when the system is already properly removed from service if all requirements of plant procedures for removing equipment from service have been met. This includes required clearance documentation, equipment and control board tagging, and properly positioned valves and power supply breakers.

(2) The second category excludes events in which an invalid ESF or RPS actuation occurs after the safety function has already been completed. (e.g., an invalid containment isolation signal while the containment isolation valves are already closed, or an invalid actuation of the RPS when all rods are full inserted).

(3) The third category excludes events in which an invalid ESF actuation occurs that involves only a limited set of ESFs (i.e., when an invalid actuation, isolation, or realignment of only the RWCU system, or any of the following ventilation systems: CREV system, reactor building ventilation system, fuel building ventilation system, auxiliary building ventilation system, or their equivalent ventilation systems, occurs). Invalid actuations that involve other ESFs not specifically excluded, (e.g., emergency core cooling system isolations or actuations; containment isolation valve closures that affect cooling systems, main steam flow, essential support systems, etc.; containment spray actuation; residual heat removal system isolations, or their equivalent systems); continue to be reportable.

Licensees continue to be required to submit LERs if a deficiency or condition associated with any of the invalid ESF actuations of the RWCU or the CREV systems (or other equivalent ventilation systems) satisfies any reportability criteria under § 50.72 and § 50.73.

Impact of the Amendments on the Industry and Government Resources

Relaxing the requirement for reporting of certain types of ESF actuations reduces the industry's reporting burden and the NRC's response burden. This reduction is consistent with the objectives and the requirements of the Paperwork Reduction Act. These amendments have no impact on the NRC's ability to fulfill its mission to ensure public health and safety because the deleted reportability requirements have little or no safety significance.

It is estimated that the changes to the existing rules will result in about 150 (or

5–10 percent) fewer Licensee Event Reports each year. Similar reductions are expected in the number of prompt event notifications reportable under 10 CFR 50.72. Some respondents, in their comments on the proposed rule, dated June 26, 1992, submitted an estimate of approximately 15 percent reduction in their reporting burden.

Summary of Comments

The NRC received 19 comments—2 from individuals, 3 from industry-supported organizations, and 14 from utilities. Except for two respondents, all commenters welcomed the Commission's efforts to reduce the licensee burden and to save the agency's resources in event review and processing. The utilities and the industry-supported organizations expressed their desire for a broader relaxation to include all invalid ESF actuations from reporting.

Other comments from the respondents concerned the following: clarification of the definition of "invalid" actuations; examples of events being exempted from reporting; consideration of plant-specific situations; exemption from reporting of the actuation of the standby gas treatment system following an invalid actuation of the reactor building ventilation system; and possibly extending relaxation of invalid actuations/isolations of RWCU from reporting to include those of the chemical and volume control system in a pressurized water reactor. The Statement of Considerations for this final rule addresses most of these concerns. Other issues and clarifications concerning event reportability will be addressed in NUREG–1022, Revision 1. However, it is not practical to address a plant-specific situation unless it relates to a generic concern.

The Commission stresses that only certain specific invalid ESF actuations are being exempted from reporting through the present amendments. NUREG–1022, Revision 1 will contain specific examples and additional guidance on events which are presently reportable as well as those which are being exempted from reporting through these amendments. In the future, the Commission will give due consideration to other proposed relaxations from event reporting after the NRC staff has had an opportunity to reassess the data needs of the agency and performed safety assessments to justify initiating a separate general rulemaking. Until such time, all events not specifically exempted in these amendments continue to be reportable.

The two respondents who opposed the proposed amendments expressed their concerns about eliminating the selected event reporting requirements. These commenters believe that the elimination of these event reporting requirements may adversely affect the NRC's information database and ultimately affect the agency's ability to carry out its mission to protect public health and safety. For many years, the NRC staff has been systematically reviewing information obtained from Licensee Event Reports. These assessments of reactor operational experience have included data on the types of events included in the three categories that the NRC is deleting from reporting. The staff's reviews and assessments of nearly 1000 reactor-years of operational experience have identified essentially no safety significance associated with the type of events included in the aforementioned three categories. The Commission has reviewed the scope of these amendments, and on the basis of the staff's assessment of the past reactor operational experience, has subsequently concluded with a reasonable confidence that relaxation from reporting of events in the three categories does not affect the agency's ability to protect public health and safety.

Based on the input from the utilities, these amendments will reduce the industry's reporting burden by about 15 percent. The estimated savings of the NRC's response burden in event review and assessment is about 5–10 percent.

Environmental Impact: Categorical Exclusion

The NRC has determined that this final rule is the type of action described in categorical exclusions 10 CFR 51.22 (c)(3)(ii) and (iii). Therefore, neither an environmental impact statement nor an environmental assessment has been prepared for this final rule.

Paperwork Reduction Act Statement

This final rule amends information collection requirements that are subject to the Paperwork Reduction Act of 1980 (44 U.S.C. 3501 et seq.). These amendments were approved by the Office of Management and Budget approval numbers 3150–0011 and 3150–0104.

Because the rule will relax existing reporting requirements, public reporting burden of information is expected to be reduced. It is estimated that about 150 fewer Licensee Event Reports (NRC Form 366) and a similarly reduced number of prompt event notifications, made pursuant to 10 CFR 50.72, will be required each year. The resulting reduction in burden is estimated to

average 50 hours per licensee response, including the time required reviewing instructions, searching existing data sources, gathering and maintaining the data needed, and reviewing the collection of information. Send comments regarding the estimated burden reduction or any other aspect of this collection of information, including suggestions for reducing this burden, to the Information and Records Management Branch (MNBB–7714), U.S. Nuclear Regulatory Commission, Washington, DC 20555; and to the Desk Officer, Office of Information and Regulatory Affairs, NEOB–3019, (3150–0011 and 3150–0104), Office of Management and Budget, Washington, DC 20503.

Regulatory Analysis

The Commission has prepared a regulatory analysis on this final rule. The analysis examines the costs and benefits of the alternatives considered by the Commission. The analysis is available for inspection in the NRC Public Document Room, 2120 L Street, NW., Lower Level, Washington, DC 20555. Single copies of the analysis may be obtained from: Raji Tripathi, Office for Analysis and Evaluation of Operational Data, U.S. Nuclear Regulatory Commission, Washington, DC 20555. Telephone (301) 492–4435.

Regulatory Flexibility Certification

In accordance with the Regulatory Flexibility Act of 1980 (5 U.S.C. 605 (B)), the Commission certifies that this rule does not have a significant economic impact on a substantial number of small entities. The final rule affects only the event reporting requirements for operational nuclear power plants. The companies that own these plants do not fall within the scope of the definition of "small entities" set forth in the Regulatory Flexibility Act or the Small Business Size Standards set out in regulations issued by the Small Business Administration Act in 13 CFR part 121.

Backfit Analysis

As required by 10 CFR 50.109, the Commission has completed an assessment of the need for Backfit Analysis for this final rule. The proposed amendments include relaxations of certain existing requirements on reporting of information to the NRC. These changes neither impose additional reporting requirements nor require modifications to the facilities or their licenses.

Accordingly, the NRC has concluded that this final rule does not constitute a backfit and, thus, a backfit analysis is not required.

List of Subjects in 10 CFR Part 50

Antitrust, Classified information, Criminal penalty, Fire prevention, Incorporation by reference, Intergovernmental relations, Nuclear power plants and reactors, Radiation protection, Reactor siting criteria, Reporting and recordkeeping.

For the reasons set out in the preamble and under the authority of the Atomic Energy Act of 1964, as amended, the Energy Reorganization Act of 1974, as amended, and 5 U.S.C. 552 and 553, the Commission is adopting the following amendments to 10 CFR part 50.

PART 50—DOMESTIC LICENSING OF PRODUCTION AND UTILIZATION FACILITIES

1. The authority citation for Part 50 is revised to read as follows:

Authority: Secs. 102, 103, 104, 105, 161, 182, 183, 186, 189, 68 Stat. 936, 937, 938, 948, 953, 954, 955, 956, as amended, sec. 234, 83 Stat. 1244, as amended (42 U.S.C. 2132, 2133, 2134, 2135, 2201, 2232, 2233, 2236, 2239, 2282); secs. 201, as amended, 202, 206, 88 Stat. 1242, as amended, 1244, 1246 (42 U.S.C. 5841, 5842, 5846).

Section 50.7 also issued under Pub. L. 95-601, sec. 10, 92 Stat. 2951 (42 U.S.C. 5851). Section 50.10 also issued under secs. 101, 185 68 Stat. 936, 955, as amended (42 U.S.C. 2131, 2235); sec. 102, Pub. L. 91-190, 83 Stat. 853 (42 U.S.C. 4332). Sections 50.13, 50.54(dd), and 50.103 also issued under sec. 108, 68 Stat. 939, as amended (42 U.S.C. 2138). Sections 50.23, 50.35, 50.55, and 50.56 also issued under sec. 185, 68 Stat. 955 (42 U.S.C. 2235). Sections 50.33a, 50.55a, and Appendix Q also issued under sec. 102, Pub. L. 91-190, 83 Stat. 853 (42 U.S.C. 4332). Sections 50.34 and 50.54 also issued under sec. 204, 88 Stat. 1245 (42 U.S.C. 5844). Sections 50.58, 50.91, and 50.92 also issued under Pub. L. 97-415, 96 Stat. 2073 (42 U.S.C. 2239). Section 50.78 also issued under sec. 122, 68 Stat. 939 (42 U.S.C. 2152). Sections 50.80-50.81 also issued under sec. 184, 68 Stat. 954, as amended (42 U.S.C. 2234). Appendix F also issued under sec. 187, 68 Stat. 955 (42 U.S.C. 2237).

For the purposes of sec. 223, 68 Stat. 958, as amended (42 U.S.C. 2273): §§ 50.5, 50.46(a) and (b), and 50.54(c) are issued under sec. 161b, 68 Stat. 948, as amended (42 U.S.C. 2201(b)); §§ 50.5, 50.7(a), 50.10(a)-(c), 50.34(a) and (e), 50.44(a)-(c), 50.46(a) and (b), 50.47(b), 50.48(a), (c), (d), and (e), 50.49(a), 50.54(a), (i), (i)(1), (l)-(n), (p), (q), (t), (v), and (y), 50.55(f), 50.55a(a), (c)-(e), (g), and (h), 50.59(c), 50.60(a), 50.62(b), 50.64(b), 50.65, and 50.80(a) and (b) are issued under sec. 161i, 68 Stat. 949, as amended (42 U.S.C. 2201(i)); and §§ 50.9(d), (h), and (j), 50.54(w), (z), (bb), (cc), and (dd), 50.55(e), 50.59(b), 50.61(b), 50.62(b), 50.70(a), 50.71(a)-(c) and (e), 50.72(a), 50.73(a) and (b), 50.74, 50.78, and 50.90 are issued under sec. 161o, 68 Stat. 950, as amended (42 U.S.C. 2201(o)).

2. In § 50.72, paragraph (b)(2)(ii) is revised to read as follows:

§ 50.72 Immediate notification requirements for operating nuclear power reactors.

* * * * *

(b) Non-emergency Events. * * *

(2) Four-hour reports. * * *

(ii) Any event or condition that results in a manual or automatic actuation of any engineered safety feature (ESF), including the reactor protection system (RPS), except when:

(A) The actuation results from and is part of a pre-planned sequence during testing or reactor operation;

(B) The actuation is invalid and:

(1) Occurs while the system is properly removed from service;

(2) Occurs after the safety function has been already completed; or

(3) Involves only the following specific ESFs or their equivalent systems:

(i) Reactor water clean-up system;

(ii) Control room emergency ventilation system;

(iii) Reactor building ventilation system;

(iv) Fuel building ventilation system; or

(v) Auxiliary building ventilation system.

* * * * *

3. In § 50.73, paragraph (a)(2)(iv) is revised to read as follows:

§ 50.73 Licensee event report system.

(a) Reportable events. * * *

(2) The licensee shall report: * * *

(iv) Any event or condition that resulted in a manual or automatic actuation of any engineered safety feature (ESF), including the reactor protection system (RPS), except when:

(A) The actuation resulted from and was part of a pre-planned sequence during testing or reactor operation;

(B) The actuation was invalid and:

(1) Occurred while the system was properly removed from service;

(2) Occurred after the safety function had been already completed; or

(3) Involved only the following specific ESFs or their equivalent systems:

(i) Reactor water clean-up system;

(ii) Control room emergency ventilation system;

(iii) Reactor building ventilation system;

(iv) Fuel building ventilation system; or

(v) Auxiliary building ventilation system.

* * * * *

Dated at Rockville, MD, this 27th day of August, 1992.

For the Nuclear Regulatory Commission.

James M. Taylor,

Executive Director for Operations.

[FR Doc. 92-21750 Filed 9-9-92; 8:45 am]

BILLING CODE 7590-01-M

FEDERAL RESERVE SYSTEM

12 CFR Part 225

[Regulation Y; Docket No. R-0706]

RIN 7100-AB09

Bank Holding Companies and Change in Bank Control

AGENCY: Board of Governors of the Federal Reserve System.

ACTION: Final rule.

SUMMARY: The Board is amending its Regulation Y to augment the list of permissible nonbanking activities for bank holding companies to include the provision of full service securities brokerage under certain conditions; and the provision of financial advisory services under certain conditions. The Board has by order previously approved these activities. Applications by bank holding companies to engage in activities included on the Regulation Y list of permissible nonbanking activities may be processed by the Reserve Banks under expedited procedures pursuant to delegated authority.

EFFECTIVE DATE: September 10, 1992.

FOR FURTHER INFORMATION CONTACT: Scott G. Alvarez, Associate General Counsel (202/452-3583), or Thomas M. Corsi, Senior Attorney (202/452-3275), Legal Division. For the hearing impaired only, Telecommunications Device for the Deaf (TDD), Dorothea Thompson (202/452-3544).

SUPPLEMENTARY INFORMATION:

Background

The Bank Holding Company Act of 1956, as amended (the "BHC Act"), generally prohibits a bank holding company from engaging in nonbanking activities or acquiring voting securities of any company that is not a bank. Section 4(c)(8) of the BHC Act provides an exception to this prohibition where the Board determines after notice and opportunity for hearing that the activities being conducted are "so closely related to banking or managing or controlling banks as to be a proper incident thereto." 12 U.S.C. 1843(c)(8). The Board is authorized to make this determination by order in an individual case or by regulation.

The Board's Regulation Y (12 CFR part 225) sets forth a list of nonbanking

NRC FORM 335
(2-89)
NRCM 1102,
3201, 3202

U.S. NUCLEAR REGULATORY COMMISSION

BIBLIOGRAPHIC DATA SHEET

(See instructions on the reverse)

1. REPORT NUMBER (Assigned by NRC, Add Vol., Supp., Rev., and Addendum Numbers, if any.)
NUREG-1022, Rev. 1

2. TITLE AND SUBTITLE

Event Reporting Guidelines
10 CFR 50.72 and 50.73
Second Draft for Comment

3. DATE REPORT PUBLISHED	
MONTH	YEAR
January	1998

4. FIN OR GRANT NUMBER

5. AUTHOR(S)

D.P. Allison, M.R. Harper, W.R. Jones, J.B. MacKinnon, S.S. Sandin

6. TYPE OF REPORT

Regulatory

7. PERIOD COVERED *(Inclusive Dates)*

8. PERFORMING ORGANIZATION - NAME AND ADDRESS *(If NRC, provide Division, Office or Region, U.S. Nuclear Regulatory Commission, and mailing address; if contractor, provide name and mailing address.)*

Office for Analysis and Evaluation of Operational Data

U.S. Nuclear Regulatory Commission

Washington, DC 20555-0001

9. SPONSORING ORGANIZATION - NAME AND ADDRESS *(If NRC, type "Same as above"; if contractor, provide NRC Division, Office or Region, U.S. Nuclear Regulatory Commission, and mailing address.)*

Same as above.

10. SUPPLEMENTARY NOTES

11. ABSTRACT *(200 words or less)*

Revision 1 to NUREG-1022 clarifies the immediate notification requirements of Title 10 of the Code of Federal Regulations, Part 50, Section 50.72 (10 CFR 50.72), and the 30-day written licensee event report (LER) requirements of 10 CFR 50.73 for nuclear power plants. This revision was initiated to improve the reporting guidelines related to 10 CFR 50.72 and 50.73 and to consolidate these guidelines into a single reference document. A first draft of this document was noticed for public comment in the Federal Register on October 7, 1991 (56 FR 50598). A second draft was noticed for comment in the Federal Register on February 7, 1994 (59 FR 5614). This document updates and supersedes NUREG-1022 and its Supplements 1 and 2 (published in September 1983, February 1984, and September 1985, respectively). It does not change the reporting requirements of 10 CFR 50.72 and 50.73.

12. KEY WORDS/DESCRIPTORS *(List words or phrases that will assist researchers in locating the report.)*

Emergency Notification System
Event Report
Immediate Notification
Licensee Event Report
Notification
Report

13. AVAILABILITY STATEMENT

unlimited

14. SECURITY CLASSIFICATION

(This Page)

unclassified

(This Report)

unclassified

15. NUMBER OF PAGES

16. PRICE

NRC FORM 335 (2-89)

www.ingramcontent.com/pod-product-compliance
Lightning Source LLC
Chambersburg PA
CBHW081446170526
45166CB00008B/2329